Growth and Processing of Electronic Materials

Also from IOM Communications

Electronic Materials for Defence and Aerospace

Materials Foresight on the Electronics Industry

Growth and Processing of Electronic Materials

Edited by

NEIL McN. ALFORD
South Bank University, UK

Book 703
First published in 1998 by
IOM Communications Ltd
1 Carlton House Terrace
London SW1Y 5DB

IOM Communications Ltd
is a wholly-owned subsidiary of
The Institute of Materials

ISBN 1-86125-072-X

Reproduced from authors' camera-ready copy
Printed and bound in the UK at
The University Press, Cambridge

Contents

Introduction

This volume comprises papers and posters given at the Materials Congress '98 in the workshop 'Growth and Processing of Materials for Electronics'. The area of materials for electronics is strong in the UK. It is far stronger than its profile within the Materials community suggests. The impetus for the Growth and Processing session came from the Electronic Applications Divisional Board of the Institute of Materials. The organising committee were Professor Arthur Willoughby, Dr Eric Yeatman, Dr Caroline Millar and Professor Neil Alford. Dr Sue Dunkerton, Professors Colin Humphreys, Peter Goodhew, Roger Whatmore and Nihal Sinnadurie made up the advisory committee. However, the success of the workshop arose from the enthusiasm of the contributors. The range of topics is both broad and of high quality. The processing includes thin film, thick film (mixed oxide and sol-gel) and bulk mixed oxide methods. Professor Don Pashley opened the session with an invited talk on epitaxy growth mechanisms. Achieving epitaxy in electronics applications is sometimes absolutely essential and this lecture demonstrated the science of the processes. The workshop covered an enormous range of materials. The materials include relaxor ferroelectrics, piezoelectrics, dielectrics, superconductors and semiconductors. The applications include infra-red detectors, dielectric resonators, micro-actuators and humidity sensors. There were interesting new structures for piezoelectrics from Birmingham University, some beautiful microscopy from Liverpool and Durham, interesting applications of electron energy loss spectroscopy from Cambridge. Dielectric resonator materials and superconductors were covered by South Bank and by Leeds who have a strong tradition in electroceramics There were papers on preparation and characterisation of both piezoelectrics and relaxor ferroelectrics. Piezoelectric thin films made by combining liquid metal precursors were described by Imperial College. Solar cell research was described by both Liverpool and Cranfield. Industrial activity on Cd–Hg–Te infra-red detectors was described by GEC Marconi Infra-Red Ltd and there was a down-to-earth paper on packaging from TWI.

The enormous current market for multilayer capacitors (a billion a day), expansion of activities requiring piezo/thermo sensors and the remarkable growth in mobile communications has resulted in an electronic ceramics market (1998) of around $5BN in the US alone (*Ceram. Bull.* **77,** p. 27 1998). UK research on growth and processing of electronic materials in university and industry is strong and interdisciplinary and inter-group links are developing in this diverse area of science.

Neil McN. Alford
South Bank University
July 1998

EPITAXY GROWTH MECHANISMS

D.W.PASHLEY
Dept of Materials, Imperial College of science, Technology and Medicine,
London SW7 2AZ, UK
Email d.w.pashley@ic.ac.uk

ABSTRACT

The mechanisms of epitaxial growth affect both the form of a deposit, and the way in which lattice defects are introduced into the deposit. This paper gives emphasis to the understanding of the growth mechanisms as determined from experimental observations, rather than theoretical treatments. The results of experimental observations discussed include those obtained by electron diffraction, electron microscopy, X-ray diffraction, scanning tunnelling microscopy and atomic force microscopy.

The structural details associated with the three well established growth modes, namely monolayer growth (Frank-van der Merwe), three-dimensional nucleation (Volmer-Weber) and monolayer growth followed by three-dimensional nucleation (Stranski-Krastanow), are discussed and summarised. Emphasis is given to the growth by molecular beam epitaxy (MBE), because it allows the very early stages to be investigated. Examples are discussed to illustrate some of the main features of the details of the growth, and how they influence the overall structure of the deposits as growth proceeds. The aspects of the structure considered include the lattice spacings of the deposit (i.e.elastic strain), its physical form (e.g. isolated islands or continuous films) and its imperfection content, especially misfit dislocations. Under some conditions, lattice tilting of the deposit occurs.

INTRODUCTION

The phenomenon of epitaxy, the oriented growth of one crystal upon another crystal, has been known for at least 170 years[1], having been observed first in minerals. Laboratory studies of epitaxy initially involved growth from solution, and optical microscopy. The most systematic studies were carried out by Royer[2], who formulated a number of rules including the requirement that a small misfit (say less than 15 %) between the crystal lattices of the substrate and the overgrowth is necessary for epitaxy to occur. Studies of epitaxial thin films formed by such techniques as condensation from the vapour phase did not commence until the 1930's, following the discovery of electron diffraction in 1927 and the subsequent introduction of the technique of reflection electron diffraction, now known as RHEED.

The RHEED technique was able to provide information on the structure and orientation of films less than one atom layer in average thickness[3], including information on the physical form of the epitaxial deposits, leading to the real possibility of gaining some knowledge of the growth mechanisms. It was possible, at that stage, to identify that growth occurred either in the form of continuous layers or by the formation of small isolated three-dimensional nuclei. It was also established that epitaxy can occur when the misfit is much greater than 15%[1]. In the late 1950's, transmission electron microscopy (TEM) had developed to the extent that it could be used in the study of epitaxial deposits, leading to a considerable increase in the scope for studies of growth mechanisms[4]. Since that time, a great deal of TEM evidence on the structure of epitaxial layers of various thicknesses has provided much information about growth mechanisms, leading to the recognition of three distinct modes of growth[5,6]. These are:

(i)	Monolayer growth	(Frank - van der Merwe)
(ii)	3D nucleation	(Volmer - Weber)
(iii)	SK mode	(Stranski - Krastanow)

The SK mode involves the growth of a small number of monolayers, followed by the occurrence of 3D nucleation.The associated names are of those who provided theoretical backgrounds to the different modes. This classification into three modes is still relevant today but, since that time, new techniques for structural studies such as high resolution X-ray diffraction (HRXRD), scanning electron microscopy (SEM), high resolution transmission electron microscopy (HRTEM), scanning tunnelling microscopy (STM) and atomic force microscopy (AFM) have added considerably to our knowledge of epitaxy. These have allowed fine-scale details of the various stages of growth, for each of the modes, to be obtained.

The purpose of this paper is to describe the various modes and mechanisms of growth which have resulted from such studies. The subject is one where the interplay between experimental studies and theoretical studies plays a key role in progress, but in this paper emphasis is given to the development of knowledge of the growth mechanisms as revealed by the application of the various techniques for structural investigation.

The growth mechanisms determine the physical form of the epitaxial deposits, in terms of continuous thin films or collections of isolated crystals. In addition, the crystal spacings of the deposits can be affected as a result of the phenomenon of pseudomorphism. The concept of pseudomorphism was introduced by Finch and Quarrell[7,8], who indicated that the spacings of an epitaxial deposit in the interface plane could be modified to match those of the substrate. Although their experimental evidence at the time was dubious[1], and still is, Frank and van der Merwe[9,10] incorporated the concept in their now classic treatment of the growth of epitaxial monolayers. Subsequent experiments have provided extensive and conclusive evidence of the occurrence of pseudomorphism. Frank and van der Merwe also introduced the concept of misfit dislocations, the formation and presence of which have been the subject of extensive experimental studies. The way in which misfit dislocations are introduced during epitaxial growth is influenced by the actual growth mechanisms.

In this paper, consideration is given to the way in which the growth mechanisms influence both the physical form of the deposits and the formation of misfit dislocations and other imperfections. The relief of epitaxial strain, mainly by the formation and movement of dislocations, and the mechanisms of growth are therefore closely linked. Thus one of the main themes presented in this paper is the inter-relationship of strain relief mechanisms and growth modes and mechanisms.

MONOLAYER GROWTH

Monolayer growth is illustrated in Figure 1. Each layer is formed by the nucleation of monolayer islands which intergrow to form a continuous complete monolayer. This process is revealed by TEM , and can be revealed and controlled during the growth of the epilayer by in-situ RHEED observations. RHEED oscillations[11] occur only for monolayer growth, and the periodicity of the oscillations allows the number of monolayers present to be determined. When monolayer islands

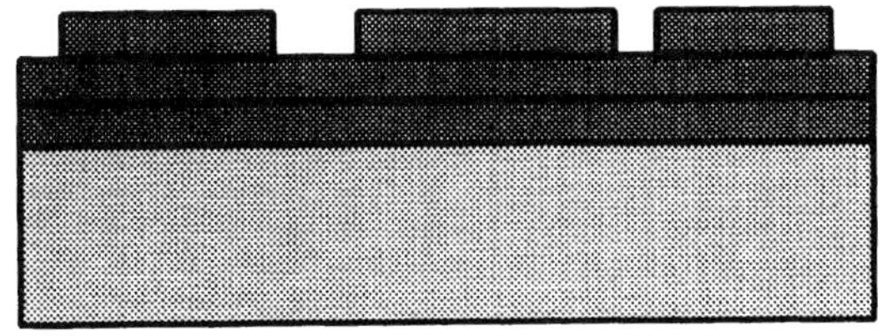

Figure 1
The monolayer growth mode

are formed initially, the theory of Frank and van der Merwe[9,10] says that the islands are pseudomorphic if the misfit is not too high, because they are of lower energy than would be the case if they had their own natural spacing. The limit will vary from one example to another, but a figure of about 10% is probably a useful guide. If the islands remain pseudomorphic as they join together to form a continuous film, there is no kind of mismatch between neighbouring islands as they coalesce, and no introduction of any defects during coalescence. Complete monlayers grow on top of each other in this way until, as the thickness increases, the pseudomorphic state is no longer favoured energetically. The layer will then start to relax, usually by plastic deformation involving the introduction of misfit dislocations, provided the stresses present in the layer are sufficient to cause both the motion and nucleation of dislocations. This is the concept of critical thickness. Much has been published on the theory and experimental measurement of critical thickness, and there are good reviews of this[12]. There are various mechanisms for introducing the misfit dislocations. Movement of any existing threading dislocations can result in the formation of misfit dislocations[13]. Otherwise, new dislocations have to be formed. They can nucleate at the surface of the epilayer and then propagate down to the interface as dislocation loops, or likewise they can nucleate at the interface and move upwards as loops. Alternatively, a dislocation source can operate within the epilayer, provided it is sufficiently thick[12]. Strain relief commences once the critical thickness is exceeded, and progresses as the epilayer growth continues, eventually resulting in the formation of a fairly regular cross-grid of misfit dislocations. The average spacing S of the dislocations is given by

$$S = 100\, b_e/m \tag{1}$$

where b_e is the edge component of the dislocations in the plane of the interface and m is the percentage misfit of the system.

It has been observed with the growth of SiGe alloys on Si(100) by the application of AFM and cross-section TEM, that surface undulations occur at thicknesses below the critical thickness[14]. The peaks of these undulations allow some elastic strain relief of the pseudomorphic strain. It is not yet clear how common is this effect.

We have to ask what is the criterion for growth by the monolayer mode? Basically, it is determined by surface energy considerations[15].

If Υ_S is the surface energy of the substrate, Υ_E is the surface energy of the epilayer and Υ_I is the interfacial energy, then monolayer islands will form in preference to 3D islands if

$$\Upsilon_E - \Upsilon_S + \Upsilon_I \leq 0 \tag{2}$$

The smaller the misfit, the lower is Υ_I likely to be which is why there is a strong tendency for low misfits between similar structures in parallel orientation to result in monolayer growth. However, other factors are involved, including the nature of the chemical bonding at the interface, so that the misfit alone is not necessarily an adequate criterion. In fact it is necessary also to take account of the elastic strain energy present in the epilayer as a result of pseudomorphism, and this strain energy increases with the increase in the number of monolayers. This is done by replacing Υ_I by Υ_{In} where Υ_{In} includes the strain energy for n layers. According to equation (2), if the substrate and epilayer materials are interchanged (i.e. B grown on A instead of A grown on B) monolayer growth should be replaced by 3D growth since $\Upsilon_E - \Upsilon_S$ cannot be negative in both cases. There does not appear to be any systematic experimental evidence to test this conclusion. Account would have to be taken of any differences in the surface structures involved, so that there is the possibility that the two surface energies are not necessarily the same for the two situations, especially when the materials are compounds (e.g. III-V compounds).

When the misfit is sufficiently low to allow pseudomorphism to occur, the misfit dislocations are introduced as summarised above. When pseudomorphism does not occur during monolayer growth, the misfit dislocations are introduced during the initial stages of growth. Two possibilities are: (i) the immediate formation of a network of dislocations between the substrate and the first monolayer, and (ii) the formation of pseudomorphic monolayer islands which become strain relieved once they exceed a certain size by the nucleation of dislocations at the edge of the islands and their movement along the interface (Figure 2). Until recently, there has been little direct experimental evidence of the processes involved, but STM is beginning to indicate that other mechanisms can occur, as shown later.

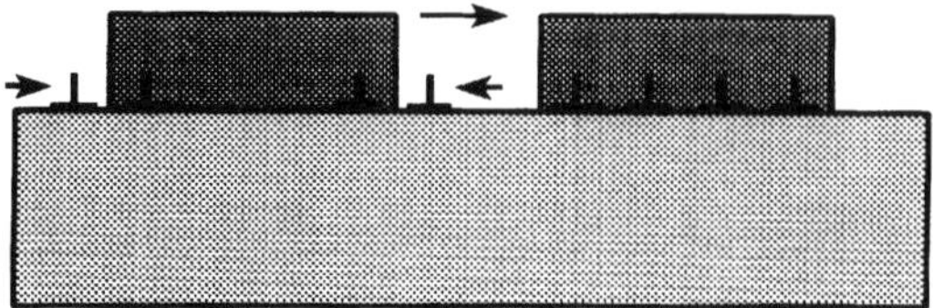

Figure 2
Dislocations feeding along the interface of the left-hand island result in the arrangement in the right-hand island

GROWTH BY 3D NUCLEI

If equation (2) is not satisfied, isolated 3D nuclei are formed initially (Figure 3).They can be crystallographic in shape, or not. The formation of nuclei obviously depends upon there being an adequately high diffusion coefficient for the migration of the deposit atoms or molecules over the substrate surface. Much has been published on the kinetices of nucleation, and the nucleation behaviour is now fairly well understood[16]. The formation of nuclei of a critical minimum size is followed by an increase in the number of nuclei as deposition continues until no further new nuclei are formed and the existing nuclei increase in size.

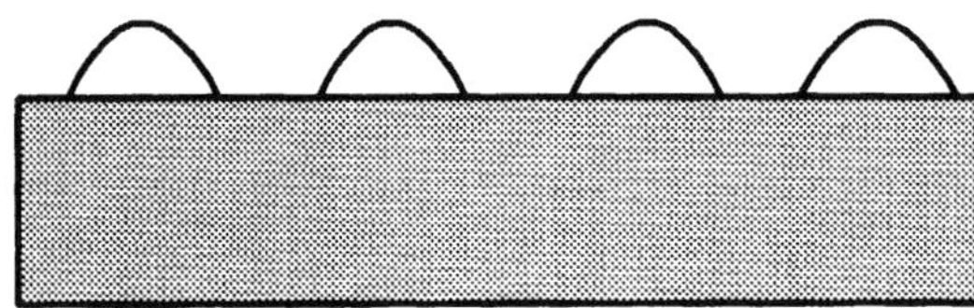

Figure 3
3D nucleation

Depending upon the misfit, the initial nuclei might or might not be pseudomorphic. If they are pseudomorphic, strain relief can take place before any coalescence occurs, by the introduction of misfit dislocations. This can happen in two different ways; basically the same mechanism as shown in figure 2 results in pure edge dislocations, or plastic deformation by the nucleation and movement of dislocations produces the so-called 60 degree misfit dislocations which are mixed in character.

Alternatively, the strain relief can occur at a later stage when coalescence has resulted in the formation of larger islands. When such relaxed nuclei or islands coalesce it is likely that additional imperfections will form because the misfit introduces a small random shift between the lattices of neighbouring islands, as indicated in Figure 4. Thus growth by the VW mechanism tends to result

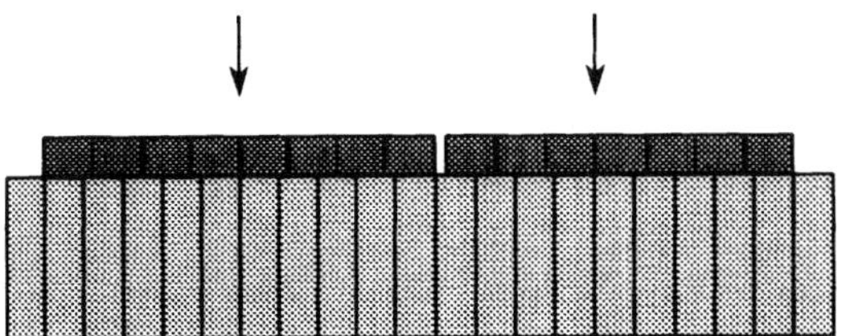

Figure 4
The shift between neighbouring Islands as a result of the lattice misfit. The arrows indicate the matching positions

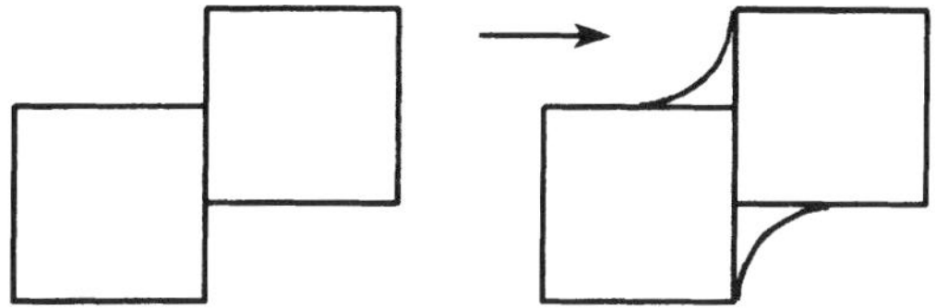

Figure 5
Selective deposition in re-entrant regions formed by coalescence

in many imperfections in the epilayer, including a high density of threading dislocations, and possibly stacking faults.

We need to consider just what happens as islands coalesce. Basically there are two possible different mechanisms involved. As neighbouring islands touch, the re-entrant regions produced are sites where it is energetically favourable for freshly arriving atoms to settle, in order to minimise surface area and hence total surface energy, so that a more equiaxed compound island is produced as deposition continues (Figure 5). This has been called selective deposition[17]. In addition, a further process can occur if there is significant surface self diffusion of the deposit material at the temperature at which deposition takes place. Surface energy minimisation results in the kind of change in shape of the compound island shown in Figure 6. Material from regions such as A is transferred to the re-entrant regions such as B, and continuation of this process results in the formation of a compound island with a shape similar to that of the two islands before

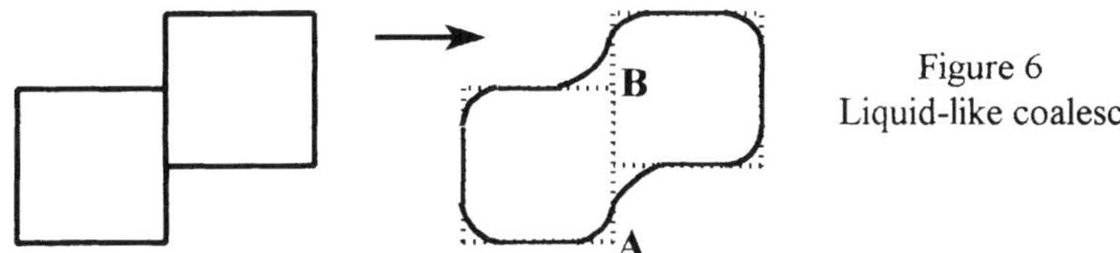

Figure 6
Liquid-like coalescence

coalescence. In extreme cases this gives a very rapid change appearing like the coalescence of liquid drops[18]. Hence the process is known as liquid-like coalescence. The effect is very pronounced with some metal deposits, although it is far less obvious for other materials, including III-V and II-VI compounds. Some of the more limited effects are described later . These two processes, especially the preferential deposition in surface hollows, continue during the later stages of growth, leading to continuity of the deposit and the elimination of holes. They can result in the formation of smooth parallel-sided epilayers following the 3D nucleation mechanism. The growth of InSb on GaAs (001), for which the misfit is about 14%, is an example of this mode[19].

THE SK GROWTH MODE

It so happens that various epilayers of semiconductors are initiated by the SK growth mode (Figure 7). It tends to happen for cases where the misfit is more than about 2%, and where the

Figure 7
The Stranski-Krastanow mode

contribution of the elastic strain energy in equation (2) is significantly large in relation to the various surface energies. The condition represented by equation (2), with Υ_l replaced by Υ_{in}, would explain why the monolayer growth becomes unfavourable energetically for such a small value of *n*. Also, for such thin layers, relief by dislocation nucleation and movement could be difficult. Well established examples are Ge and SiGe alloys on Si(001), and InAs and InGaAs alloys on GaAs(001). Typically, growth commences by the deposition of a few continuous monolayers, usually between one and ten, forming what is commonly known as the wetting layer. This is then followed by the formation of small 3D nuclei on top of the wetting layer. Initial evidence suggested that the 3D nuclei are strain relaxed by having a network of misfit dislocations at their interface with the wetting layer, so that it is energetically favourable for the nuclei to form rather than for the strained monolayer growth to continue. Although this behaviour appears to apply to some systems, some more recent evidence has shown that the earliest 3D nuclei are pseudomorphic in the two systems mentioned above[20,21], at least to the extent that there is coherence at the interface. There is no widely accepted explanation as to why the SK growth mode applies under these conditions. It is commonly explained in terms of some limited strain relief which means that the nucleation stage is a lower energy stage than continued growth of pseudomorphic monolayers. Some strain relief can be obtained by local elastic deformation of the wetting layer and the substrate[22], as illustrated schematically in Figure 8. Because the small nuclei are not constrained as are the monolayers, they are able to deform elastically in directions parallel to the interface (see Figure 9), also giving some relaxation of the epitaxial strain. The formation

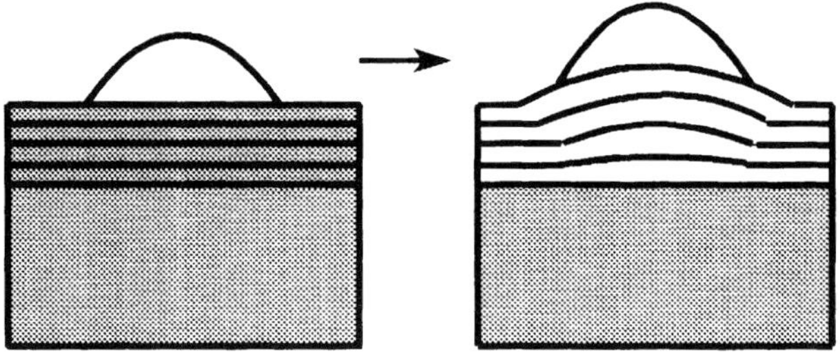

Figure 8
Partial pseudomorphic strain relief by elastic strain of the substrate (see reference 22)

of small pseudomorphic nuclei during SK growth has become of serious interest because it provides a possible route for the production of arrays of quantum dots for device applications. The early stages of growth can now be investigated in much more detail by STM. If MBE growth takes place in a high vacuum chamber from which the sample can be transferred to the STM chamber without any exposure to air or other higher pressure gases, there is no damage of the deposit before it is examined. It has been found that for growth of Ge on Si(001) in the

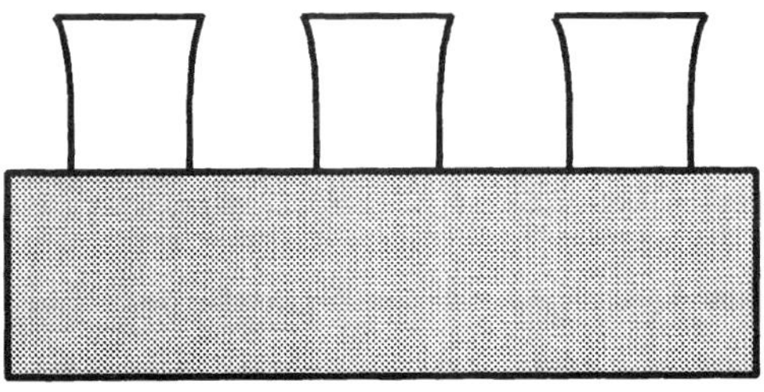

Figure 9
Schematic representation of elastic strain of pseudomorphic nuclei for positive misfit

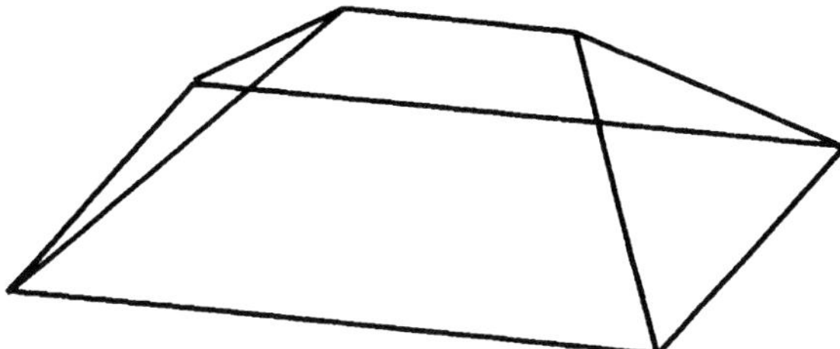

Figure 10
The form of the hut clusters observed for Ge on Si(001). Typical height is 30Å. The lengths range from about 200 Å upwards

region of 500°C, the initial formation of three monolayers is followed by the nucleation of small crystallographically shaped islands which have become known as hut clusters[23,24] (see Figure 10).These roof-top shaped islands remain pseudomorphic, and strain relief occurs only after the nucleation of a new crop of more equi-axed islands after the substrate has been more or less covered by the hut clusters. These new islands are also pseudomorphic initially, and strain relief occurs after they have coalesced into bigger islands. It appears that their size range can be very uniform. This is not fully understood, but has been explained by the effect of the misfit. As a pseudomorphic nucleus grows, the strain around the edge of the nucleus increases so that it is less favourable energetically for further deposition to occur on the larger nuclei rather than the smaller nuclei. Hence, if there is a distribution of sizes, deposition occurs preferentially on the smaller nuclei causing them to grow more rapidly than the larger ones, thus leading to a smaller spread in the range of sizes[25].

The early stages of growth can now be investigated in further detail by in-situ STM.In the case of Ge grown on Si(001) by gas source MBE, this has allowed the deposition to be carried out whilst STM observations are being made[26] , so that the growth process is observed directly. This is a powerful technique since it reveals the structure on the same area of the sample as growth proceeds, although the time to form scanning images means that changes cannot be observed over time scales of much less than a minute. The development of the surface coverage by hut clusters is revealed. The distribution of sizes is analysed at the various stages and the way in which coarsening, i.e. the reduction in the number of clusters as their average size increases, takes place is revealed. It occurs by coalescence, not dynamic coalescence involving movement of nuclei over the surface, but by static coalescence when two closely neighbouring clusters enlarge and touch (see Figure 5 of reference 26). There is a shape change by 'fast diffusive interaction' (see Figure 11). Material is transferred from the region B to the region A. This appears to be the same basic process as illustrated in Figure (6), called 'liquid-like coalescence', due to self diffusion of the Ge over the surfaces of the compound island. The rate of change is far slower than the earlier

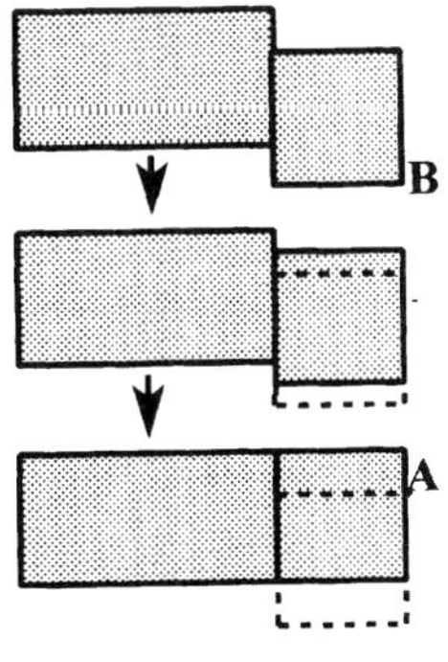

Figure 11
The static coalescence of Ge hut clusters, involving the transfer of Ge from the region B to the region A by surface diffusion

observed examples of gold and silver[1], due no doubt to a significantly lower surface self-diffusion coefficient.In-situ TEM growth has also been used to study the growth of Ge on Si(001) formed by the decomposition of germane[27]. This allows changes over much shorter timescales to be observed. As already established, the growth starts by the formation of three monolayers of Ge, mainly by growth at surface steps on the thin film substrates. Changes in the nature of the reconstructed surface took place during this stage. The subsequent stages were temperature dependent. At about 600°C pseudomorphic islands formed on top of the monolayers, and grew to about 1000Å in diameter before they started to undergo strain relief. This came about by the introduction of 60° misfit dislocations, involving slip on the inclined {111} planes. As each dislocation appeared, the islands rapidly increased in area, which was interpreted as a shape change (i.e. an accompanying reduction in height). For lower substrate temperatures at which dislocations are no longer mobile, pseudomorphic hut clusters are first formed and then new small roughly equi-axed pseudomorphic islands are nucleated. Strain relaxation of these cannot occur by slip, because dislocations are not mobile. Instead, 90° misfit dislocations form at the edges of the islands and as each one is formed the island rapidly increases in lateral size by a certain amount so that the outer regions of the growing island is largely strain relieved whilst the central region remains strained. Thus there is a clear difference between the strain relief mechanisms for different substrate temperatures, in this case dependent upon whether or not dislocations are mobile.

Both of the above in-situ techniques are clearly very powerful and informative, providing valuable detailed information on growth mechanisms and strain relief. However, it is always necessary to exercise some caution in the interpretation of the observations made with these in-situ techniques because some effects could be influenced by the technique employed (e.g. electron beam effects or specific thin film effects). It is necessary, as far as possible, to check that the structures observed at the different stages are the same as those found when growth occurs in more conventional apparatus. The great advantage of the in-situ techniques is their ability to reveal how a sequence of changes comes about, and there seems little doubt that in the examples just quoted the basic processes described above are real.

The growth of InAs and InGaAs alloys on GaAs(001) also occurs by the SK mode, and various studies of the small 3D nuclei have been made. The occurrence of the SK mode is temperature dependent and for $In_{0.5}Ga_{0.5}As$, for example[28], SK growth occurs at 520°C with 3D nuclei forming after the growth of about 5 monolayers whilst continued monolayer growth occurs at 320°C. This is considered to be due to differences in surface diffusion[28]. At 420°C InAs forms 3D nuclei after the deposition of about 1.7 monolayers[29]. It is not clear why a non-integral number of monolayers applies. There is the complication that significant interdiffusion or alloying occurs between the InAs and the GaAs, so that the immediate surface composition is uncertain and this could effect the actual surface coverage. No equivalent of the hut clusters is found, and the initial pseudomorphic nuclei are roughly equi-axed. STM observations[29] show that many of the initial 3D islands are nucleated at steps present on the wetting layer and that the increase in numbers occurs with little change in average size as growth proceeds. They have a very uniform size distribution[29,30], as already mentioned for the case of germanium on silicon.

Another important observation is that the composition of the nuclei is far from pure InAs; they contain significant amounts of GaAs, which must come about by diffusion from the substrate. Once SK nuclei grow and coalesce, the subsequent growth mechanisms are much the same as those which occur for the VW growth mode.

EFFECT OF VARYING THE ORIENTATION OF THE SUBSTRATE

The majority of the studies of the epitaxial growth of semiconducting materials, including silicon, germanium, III-V compounds and II-VI compounds have been carried out with substrates in (001) orientation. The substrate orientation can have an influence on the mode of growth of an epilayer, partly because surface energies and interface energies can be different. InAs on GaAs is an example. SK growth occurs on (001) substrates, but monolayer growth is found on (110) and (111)A substrates for the identical conditions of growth[29]. In addition, small angular deviations from low index substrate surfaces can result in a significant change in the mode of growth. Such surfaces are known as vicinal surfaces, which must contain surface steps. These steps can act as preferential sites for the trapping of deposit atoms, so that nuclei form preferentially on the steps. Growth of InAs on vicinal (110) GaAs surfaces provides a good example of the effect[31]. Clearly, there must be sufficient surface mobility of the deposit atoms and the trapping efficiency of the steps must be sufficiently high to avoid nucleation between the steps. In this case, the steps on the substrate surface are not just one monolayer in height because step bunching has occurred during the growth of the usual buffer layer. As a result, the nuclei which form on the steps are 3D rather than 2D and the surface coverage is therefore much less than for monolayer growth. The form of the deposit is very different from that which occurs on an exact (110) surface, where monolayer growth takes place.

The orientation of the substrate also influences the detailed strain relief mechanisms. For substrates in (001) orientation there are ample slip systems to allow strain relief in all directions in the interface. The same applies to substrates in (111) orientation. In both of these cases, the symmetry of the arrangements of the slip planes allows the strain relief to be isotropic, although this is not always exactly so. However, for some other substrate orientations the normal slip systems are not favourable for easily relieving strain equally in all directions. For a misfit dislocation produced along the intersection of its slip plane with the interface, and having an edge component of b_e in the interface, the displacement contribution to strain relief along a direction at an angle θ to the dislocation line is given by $b_e sin\theta$. This, when coupled with a lower symmetry of slip plane intersections, results in residual asymmetry in strain relief for all possible combinations of misfit dislocation distribution. The unique limiting case applies to (110) substrates, because the normal slip systems do not allow any strain relief in the $[1\bar{1}0]$ direction in the interface. This occurs because two of the four {111} slip planes are normal to the surface, and cannot therefore contribute to strain relief, whilst both of the other two slip planes intersect the surface along the $[1\bar{1}0]$ direction (Figure 12) and thus relieve strain along [001] but not along $[1\bar{1}0]$. For growth of InAs on (110) GaAs, with a misfit of 7%, this presents no problem because pure edge misfit dislocations, with Burgers vectors of ½a$[1\bar{1}0]$, form during the very early stages (details given below) relieving strain along $[1\bar{1}0]$[32]. Strain relief along [001] occurs by slip at a later stage. However for $In_xGa_{1-x}As$ alloys, with x<0.25 and a misfit of below 2%, the pure edge misfit dislocations are not formed in the early stages[33]. As a result, strain relief along $[1\bar{1}0]$ is

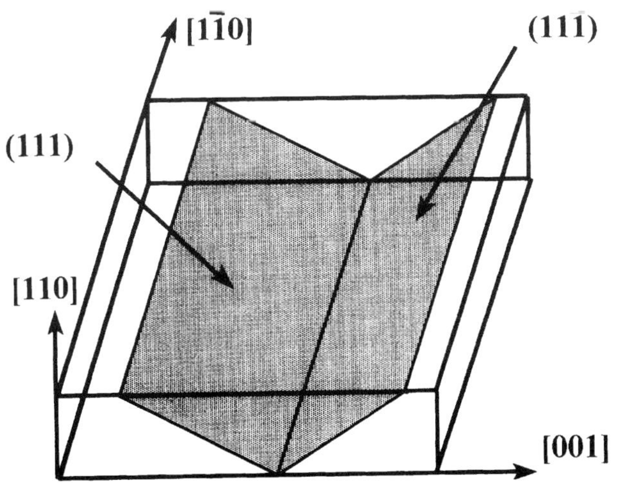

Figure 12
The operative slip planes in (110) layers

inhibited, and cannot occur by normal slip processes. Linear elastic strain remains for thicknesses well above the normal critical thickness, and occurs eventually partially by slip on {311} planes and then by a mechanism involving the formation of growth twins. These are not deformation twins, which anyway cannot relieve strain along [1$\bar{1}$0].

MONOLAYER GROWTH - LARGER MISFIT

The early stages of monolayer growth can now also be investigated in much more detail by STM. This is allowing more information to be obtained where the misfit is sufficiently high that misfit dislocations are formed in the initial stages of growth. It is particularly useful that interfacial misfit dislocations can be revealed in the STM, because of the effect of their strain field on the surface profile of the epilayer[34]. For a negative misfit, a surface ridge is formed, and for a positive misfit a surface trough occurs. For the case of InAs on GaAs (111)A growth occurs by the formation of clusters of small monolayer islands which are approximately circular in shape and which do not coalesce initially[35]. The lateral dimensions of the islands are in the range 50-100Å. Further deposition results in the formation of dislocations at the junctions between neighbouring islands as they coalesce (see A on Figure 13), and these develop into a hexagonal-type network of misfit dislocations at the interface. There are open nodes which represent triangular areas of stacking fault. The surface troughs are only about 0.5Å deep. It appears that the lateral dimensions of the islands are limited by the misfit, so that they can remain coherent with the substrate up to a certain critical size, beyond which misfit dislocations have to be introduced.

The mechanism for growth on a (110) substrate is rather different[36]. Initially, growth occurs by monolayer island formation both on large areas in between substrate surface steps and on the steps themselves. From about 0.75ML onwards cracks or fractures start to appear in the extensive monolayer islands (see Figure 14). They seem to start at the edges of islands and propagate inwards so that the deposit develops a sort of crazy paving appearance. As deposition continues, with the outer InAs layers maintaining the same structure, changes start to take place between 3

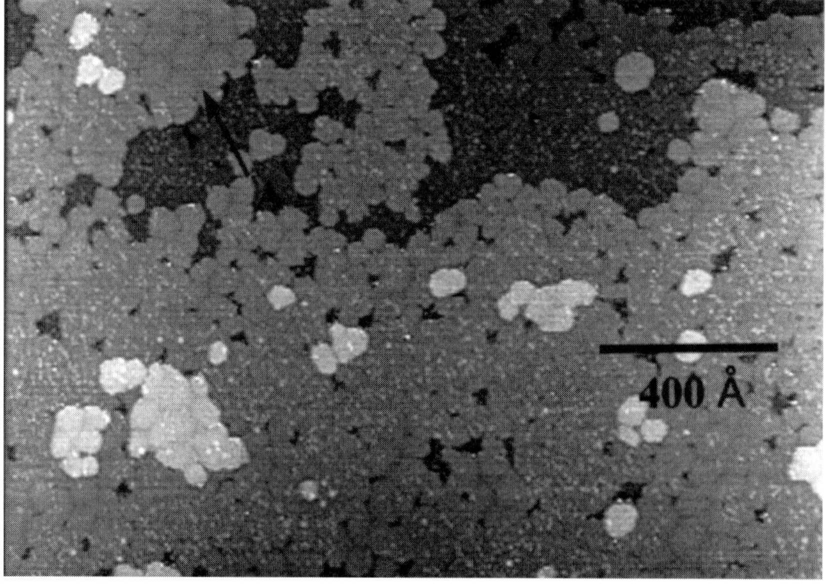

Figure 13
Filled states STM image of 2ML of InAs grown on GaAs (111)A

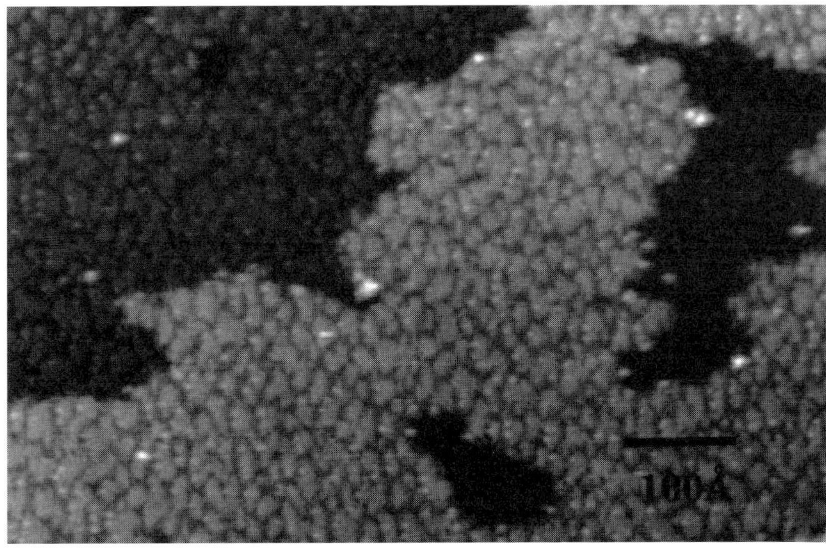

Figure 14
Filled states STM image of 2ML of InAs grown on GaAs (110)

and 5 ML in thickness. Coalescence leads to the elimination of the crazy paving structure, and misfit dislocations start to appear in the coalesced regions (see the dark horizontal lines on Figure 15). They do not form at all of the boundaries of the islands and are linear, approximately along [001], not relieving any strain along [1$\bar{1}$0]. The dislocations appear to form not precisely at the epilayer interface, but two or three layers away form it. Confirmation of the dislocation structure is given by comparing TEM images with STM images from similar samples. The dislocations are the pure edge dislocations with Burgers vectors lying in the (110) plane.

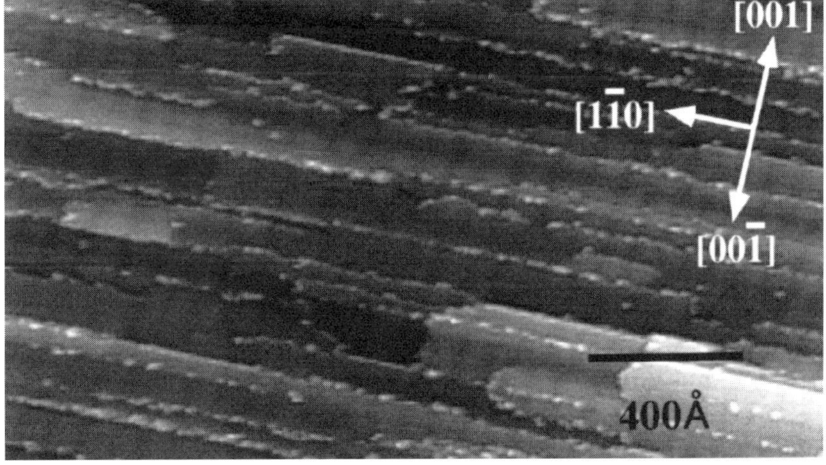

Figure 15
Filled states STM image of 4ML of InAs on GaAs (110).

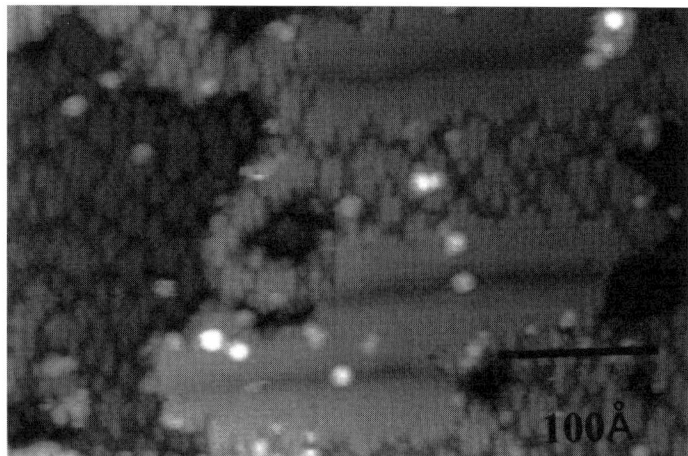

Figure 16
Filled states STM of 10ML of InAs grown on GaAs (110) showing slip steps along $[1\bar{1}0]$

Relief of strain along [001] occurs after further growth, as observed by TEM[32], by slip on the inclined (111) and $(11\bar{1})$ planes. This is revealed by the slip steps formed on the epilayer surface (see Figure 16); the edge dislocations are visible as the dark lines along $[1\bar{1}0]$.

TILTING OF EPILAYERS

The occurrence of tilting of epilayers has been observed in several systems, mainly other than semiconductors[37], and occurs in a clear way on (110) substrates[32] because there is only one possible axis of tilt. The simplest explanation is that slip on just one set of planes causes the tilting, with the misfit dislocations forming a low angle boundary[38] (see figure 12). The angle of tilt, α, is then given, to a first approximation, by

$$\tan \alpha = m \tan \phi / 100 \qquad (3)$$

where m% is the misfit and ϕ is the angle between the operating slip plane and the interface. In practice, slip is commonly distributed between the two available slip planes, in the case of (110) substrates, so that the direction of tilting varies in sign from one area to another. Tilting in one direction over a large area can be detected and measured by X-ray diffraction, but selected area diffraction in the TEM is required to reveal the tilting confined to small areas.

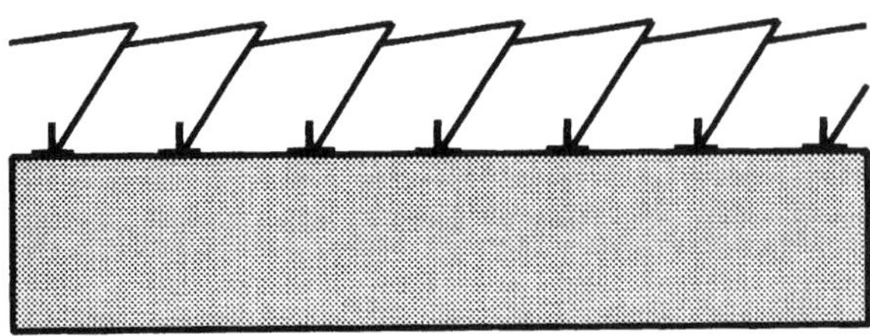

Figure 17
The formation of a tilt boundary by slip on a sequence of parallel planes

SUMMARY

The characteristics of the different modes of growth can be summarised in the following way:

MONOLAYER GROWTH
Low misfit. Pseudomorphic growth for thicknesses of at least several monolayers, many for really small misfits, followed by subsequent strain relief by slip processes.
Higher misfit. Pseudomorphic growth of monolayer islands with some, or all, (pure edge) misfit dislocations forming during the formation of the first few monolayers. The details of this kind of behaviour is likely to depend upon the orientation of the substrate surface The edge dislocations which are formed are not necessarily exactly at the interface. Subsequently, strain relief will occur by slip processes, where complete strain relief has not occurred during the initial stages of growth .

3D NUCLEATION
Low misfit. Initial nuclei are pseudomorphic and islands formed by coalescence of nuclei remain pseudomorphic. Relaxation of larger islands occurs by either slip (60° MD's), or the formation of pure edge MD's at the interface.
Higher misfit. Initial nuclei are not pseudomorphic.
Coalescence and further growth can lead to the formation of layers of uniform thickness, but a high density of threading dislocations is likely.

SK GROWTH
Initially a few pseudomorphic monolayers are formed, the higher the misfit the smaller the number of these monolayers. This is followed by the formation of small 3D nuclei, which can be pseudomorphic and of a fairly uniform size distribution. Islands formed by coalescence eventually relax by either slip or the formation of pure edge misfit dislocations at the interface. Further growth proceeds as for 3D nucleation.

These conclusions are based upon studies of a limited number of systems, which are mainly ones involving a positive misfit and compressive strains in the epilayer. More detailed studies on other systems and the possibility of new or improved techniques for studying all systems is likely to result in further kinds of behaviour being identified, and further understanding of the behaviour which has already been revealed

ACKNOWLEDGEMENTS

Discussions with Professor B.A.Joyce have been particularly helpful, and Dr. J.A.Belk is thanked for providing Figures 13,14,15 and 16.

REFERENCES

1. D.W.PASHLEY: 'Epitaxy in thin surface films', *Adv.Phys.*, 1956, **5**,173-240.
2. L.ROYER: 'Recherches expérimentales sur l'épitaxie ou orientation mutuelle de cristaux d'espèces différentes', *Bull. Soc. Franç. Min.*, 1928, **51**, 7-159.
3. R.C.NEWMAN and D.W.PASHLEY: 'The sensitivity of electron diffraction as a means of detecting thin surface films', *Phil. Mag.*, 1954, **46**, 927-940.
4. D.W.PASHLEY: 'The nucleation, growth, structure and epitaxy of thin surface films', *Adv.Phys.*, 1965, **14**,327-416.
5. E.BAUER: 'Phenomenological theory of crystal precipitation on surfaces. I', *Z.Kristallogr.*, 1958, **110**, 372-394.

6. E.BAUER and H.POPPA: 'Recent advances in epitaxy', *Thin Solid Films*, 1972, **12**, 167-185.
7. G.I.FINCH and A.G.QUARRELL: 'The structure of magnesium, zinc and aluminium films', *Proc. Roy. Soc. A*, 1933, **141**, 398-414.
8. G.I.FINCH and A.G.QUARRELL: 'Crystal structure and orientation in zinc-oxide films', *Proc. Phys. Soc.Lond.*, 1934, **46**, 148-162.
9. F.C.FRANK and J.H.VAN DER MERWE: 'One-dimensional dislocations. I. Static theory', *Proc. Roy. Soc. A*, 1949, **198**, 207-216.
10. F.C.FRANK and J.H.VAN DER MERWE: 'One-dimensional dislocations. II. Misfitting monolayers and oriented overgrowth', *Proc. Roy. Soc. A*, 1949, **198**, 216-225.
11. J.H.NEAVE, B.A.JOYCE, P.J.DOBSON and N.NORTON: 'Dynamics of film growth of GaAs by MBE from RHEED observations', *Appl. Phys. A*, 1984, **31**, 1-8.
12. R.BEANLAND, D.J.DUNSTAN and P.J.GOODHEW: 'Plastic relexation and relaxed buffer layers for semiconductor epitaxy', *Adv.Phys.*, 1996, **45**, 87-146.
13. J.W.MATTHEWS: 'Coherent interfaces and misfit dislocations', *Epitaxial Growth*, J.W.Matthews ed., Academic Press, New York, 1975, 560-609.
14. D.DUTARTRE, P.WARREN, F.CHOLLET, F.GISBERT, M.BÉRENGUER and I.BERBÉZIER: 'Defect-free Stranski-Krastanov growth of strained $Si_{1-x}Ge_x$ layers on Si', *J.Cryst. Growth*, 1994, **142**, 78-86.
15. R. KERN, G. LE LAY and J.J.METOIS: 'Basic mechanisms in the early stages of epitaxy', *Current Topics in Materials Science'*, 1979, **3**, 131-419.
16. J.A.VENABLES, G.D.T.SPILLER and M.HANDBŰCHEN: 'Nucleation and growth of thin films', *Rep. Progr. Phys.*, 1984, **47**, 399-459.
17. D.W.PASHLEY: 'The epitaxy of metals', *Materials Science and Technology*, R.W.Cahn ed., VCH, Weinheim, New York, Basle, Cambridge, 1991, **15**, 289-328.
18. D.W.PASHLEY, M.J.STOWELL, M.H.JACOBS and T.J.LAW: 'The growth and structure of silver deposits formed by evaporation inside an electron microscope', *Phil. Mag.*, 1964, **10**, 127-158.
19. X.ZHANG, A.E.STATON-BEVAN, D.W.PASHLEY, S.D.PARKER, R.DROOPAD, R.L.WILLIAMS and R.C.NEWMAN: 'A transmission electron microscopy and reflection high-energy electron diffraction study of the initial stages of heteroepitaxial growth of InSb on GaAs(001) by molecular beam epitaxy', *J. Appl. Phys.*, 1990, **67**, 800-806.
20. M.TOMITORI, K.WATANABE, M.KOBAYASHI and O.NISHIKAWA: 'STM study of the Ge growth mode on Si(001) substrates', *Appl. Surf. Sci.*, 1994, **76/77**, 322-328.
21. C.W.SNYDER, B.G.ORR, D.KESSLER and L.M.SANDER: 'Effect of strain on surface morphology in highly strained InGaAs films', *Phys.Rev.*, 1991, **66**, 3032-3035.
22. D.J.EAGLESHAM and M.CERULLO: 'Dislocation-free Stranski-Krastanow growth of Ge on Si(100)', *Phys. Rev. Lett.*, 1990, **64**, 1943-1946.
23. Y.-MO, D.E.SAVAGE, B.S.SWARTZENTRUBER and M.G.LAGALLY: 'Kinetic pathways in Stranski-Krastanov growth of Ge on Si(001)', *Phys. Rev. Lett.*, 1990, **65**, 1020-1023.
24. J.KNALL and J.B.PETHICA: 'Growth of Ge on Si(100) and Si(113) studied by STM', *Surf. Sci.*, 1992, **265**, 156-167.
25. Y.CHEN and J.WASHBURN: 'Structural transition in large-lattice-mismatch heteroepitaxy', *Phys. Rev. Lett.*, 1996, 77, 4046-4049.
26. I.GOLDFARB, P.T.HAYDEN, J.H.G.OWEN and G.A.D.BRIGGS: 'Competing growth mechanisms of Ge/Si(001) coherent clusters', *Phys. Rev. B*, 1997, **56**, 10459-10468.
27. M.HAMMAR, F.K.LEGOUES, J.TERSOFF, M.C.REUTER and R.M.TROMP: 'In situ ultrahigh vacuum transmission electron microscopy studies of hetero-epitaxial growth I.Si(001)/Ge', *Surf. Sci.*, 1996, **349**, 129-144.
28. C.W.SNYDER, J.F.MANSFIELD and B.G.ORR: 'Kinetically controlled critical thickness for coherent islanding and thick highly strained pseudomorphic films of $In_xGa_{1-x}As$ on GaAs(100)' *Phys. Rev. B*, 1992, **46**, 9551-9554.

29. B.A.JOYCE, J.L.SUDIJONO, J.G.BELK, H.YAMAGUCHI, X.M.ZHANG, H.T.DOBBS, A.ZANGWILL, D.D.VVEDENSKY and T.S.JONES: ' A scanning tunnelling microscopy-reflection high energy electron diffraction-rate equation study of the molecular beam epitaxial growth of InAs on GaAs(001), (110) and (111)A - quantum dots' *Jpn. J. Appl. Phys.*, 1997, **36**, 4111-4117.

30. J.M.MOISON, F.HOUZAY, F.BARTHE, L.LEPRINCE, E.ANDRÉ and O.VATEL: 'Self-organised growth of regular nanometer-scale InAs dots on GaAs', *Appl. Phys. Lett.*, **64**,196-198.

31. X.M.ZHANG, D.W.PASHLEY, I.KAMIYA, J.H.NEAVE and B.A.JOYCE: 'The nucleation and growth by molecular beam epitaxy of InAs on GaAs(110) misoriented substrates', *J. Cryst. Growth*, 1995, **147**, 234-237.

32. X.ZHANG, D.W.PASHLEY, L.HART, J.H.NEAVE, P.N.FAWCETT and B.A.JOYCE: 'The morphology and asymmetric strain relief behaviour of InAs films on GaAs (110) grown by molecular beam epitaxy', *J. Cryst. Growth*, 1993, **131**, 300-308.

33. X.ZHANG, D.W.PASHLEY, J.H.NEAVE, L.HART and B.A.JOYCE: 'The strain relaxation of $In_{0.1}Ga_{0.9}As$ on GaAs (110) grown by molecular beam epitaxy', *J. Appl. Phys.*, 1995, **78**, 6454-6457.

34. J.G.BELK, L.SUDIJONO, X.M.ZHANG, J.H.NEAVE, T.S.JONES and B.A.JOYCE: 'Surface contrast in two dimensionally nucleated misfit dislocations in InAs/GaAs(110) heteroepitaxy', *Phys. Rev. Lett.*, 1997, **78**, 475-478.

35.H.YAMAGUCHI, J.G.BELK, X.M.ZHANG, J.L.SUDIJONO, M.R.FAHY, T.S.JONES, D.W.PASHLEY and B.A.JOYCE: 'Atomic-scale imaging of strain relaxation via misfit dislocations in highly mismatched semiconductor heteroepitaxy: InAs/GaAs(111)A', *Phys. Rev. B*, **55**, 1337-1340.

36. J.G.BELK, D.W.PASHLEY, C.F.McCONVILLE, J.L.SUDIJONO, B.A.JOYCE and T.S.JONES: 'Surface atomic configurations due to dislocation activity in InAs/GaAs(110) heteroepitaxy', *Phys. Rev. B*, 1997, **56**, 10289-10296.

37. R.DU and C.P.FLYNN: 'Asymmetric coherent tilt boundaries formed by molecular beam epitaxy', *J. Phys. Condens. Matter*, 1990, **2**, 1335-1341.

38. D.W.PASHLEY: 'The relief of pseudomorphic strain in epilayers of f.c.c. structures grown in (110) orientation', *Phil. Mag. A*, 1993, **67**, 1333-1346.

DEFECT STUDIES IN SULPHUR-DOPED InGaAs/GaAs GROWN BY CHEMICAL BEAM EPITAXY

G.M. PETKOS*, P.J. GOODHEW**
Materials Science and Engineering, University of Liverpool, Liverpool L69 3BX
* Email petkos@liverpool.ac.uk
** Email goodhew@liverpool.ac.uk

ABSTRACT

Sulphur-doped $In_xGa_{1-x}As$ layers have been grown by chemical beam epitaxy (CBE) on GaAs substrates, and characterised using secondary ion mass spectrometry (SIMS), capacitance-voltage (C-V) profiling, and transmission electron microscopy (TEM).
Diethylsulphide (DES) was used successfully as the doping precursor. Carrier concentrations as high as $10^{20}cm^{-3}$ were measured, limited not by the electrical activation but by the precursor flux.
The carrier concentration rises or drops sharply near the substrate-epilayer interface and also near the interfaces between the epilayers, depending on the doping level. The decrease in carrier concentration near the interface was attributed to the increased recombination due to misfit dislocations, whereas the increase in the case of lower doping levels appears to be an artefact due to the etching of the InGaAs/GaAs or GaAs/GaAs interface.
TEM observations revealed unusual dislocation configurations such as non-straight dislocation lines and gamma-shaped dislocation loops, not seen in undoped InGaAs/GaAs. Dislocation climb could account for the development of such configurations via point defect absorption or emission.

INTRODUCTION

InGaAs has been widely studied recently in relation to several heterostructure device applications. Having a higher carrier mobility and saturation velocity than GaAs, it is suitable for applications in high frequency microelectronics [1] and optoelectronic devices [2].
Sulphur, with its low diffusivity in GaAs [3] is an attractive n-type dopant for III-V semiconductors grown by chemical beam epitaxy (CBE). Hirtz [3] demonstrated that S-doped GaAs can be grown successfully by CBE, using hydrogen sulphide (H_2S) as the sulphur precursor. However, as H_2S is highly toxic, alternative precursors have been studied. Joyce et al [4] investigated the use of four metallorganic sulphur sources for the doping of GaAs, concluding that diethylsulphide (DES) is the most appropriate. Consequently, DES was used to study S-doping of GaAs, AlGaAs and InGaAs [5], demonstrating that it is an attractive replacement for H_2S in the CBE growth of these compounds.
During the present study, DES was used for the doping of InGaAs/GaAs layers grown by CBE. In addition to the doping incorporation and carrier concentration studies, microstructural investigations were carried out. The dislocation behaviour was found to be dependent on the sulphur concentration levels. The study of relaxation mechanisms in relation to the interaction between impurities and dislocations is of great importance for the InGaAs/GaAs system, as misfit dislocations can cause severe degradation of the heteroepitaxial devices.

EXPERIMENT

A VG semicon V80H gas source growth system was used to grow S-doped InGaAs/GaAs layers. The gas lines utilise pressure control of flux through a fine orifice with 0-10 Torr baratrons on the metallorganic source lines. A detailed description of the system is given by Joyce [6]. InGaAs layers were grown on both n-type (Si-doped) and semi-insulating GaAs(001) wafers with a GaAs buffer layer between the InGaAs layer and the GaAs substrate. The growth rate was 1μm/h. Triethylgallium and trimethylindium were used as group III sources. All InGaAs samples were grown at a substrate temperature of 500°C.
SIMS profiling was carried out using a 10keV Cs^+ ion beam on a Cameca Ims 3f system calibrated against known standards. Electrochemical C-V profiling measurements were carried out using a Bio-Rad PN4300 semiconductor profile plotter.
TEM studies were carried out using a Phillips EM 400T microscope operating at 120kV and a JEOL 2000 FX microscope operating at 200 kV. Scanning transmission electron microscopy (STEM) studies were carried out using a VG HB601 microscope operating at 100kV. **g** <220> reflections were used in order to image 60° misfit dislocations.

RESULTS

Sulphur incorporation and electrical activation

Sulphur concentration [S] profiles obtained by SIMS for various layers of $In_{0.2}Ga_{0.8}As$ indicate that [S] is limited by the available DES flux. No evidence of saturation was observed up to the range of 10^{20} cm^{-3}, as it is demonstrated in figure 5.1, where SIMS results from sample 1 are presented. During the growth of this sample, the DES flux was increased from 7.5 Torr to the maximum available flux of 10 Torr, resulting in the increase in [S], shown in figure 1. The nominal layer thickness, sulphur concentration, and In composition of the samples used are presented on Table 1.

	h(nm)	P (Torr)	Inx	[S] (cm-3)
Sample 1	1000/	10/7.5	0.2	$8/3x10^{19}$
	250 buffer	0	0	$1x10^{17}$
Sample 2	50/	0	0	-
	1000/	0.5	0.2	$8x10^{17}$
	500 buffer	0.04	0	$5x10^{17}$
Sample 3	250/	10	0.2	$1.5x10^{20}$
	250 buffer	0	0	$1x10^{17}$

Table 1: Layer thickness h, DES baratron pressure P, In composition x, and sulphur concentration [S] of the samples 1, 2 and 3.

Comparison of sulphur concentration [S] and carrier concentration N results, measured by SIMS and C-V profiling respectively, indicates that the activation efficiency is high, reaching 100% in some cases. Also, no evidence of saturation of the electrical activation was found. These results demonstrate that DES is a promising precursor for S-doping of semiconductors grown by CBE, as it is less toxic than H_2S and it produces high carrier concentrations.

The carrier concentration decreases near the InGaAs/GaAs interface as shown in figure 2. In the case of relatively low doping levels ($N<5x10^{18}$ cm^{-3}), however, an increase appears near the interface, in the shape of a spike on the C-V profile similar to the one shown in figure 2.

The decrease of carrier concentration near the interface has been reported before, in the case of heavily Si-doped MBE-grown $In_{0.5}Ga_{0.5}As$/GaAs [7]. Both carrier concentration and electron mobility were found to decrease towards the interface, and this behaviour was attributed to the presence of misfit dislocations, which act as both recombination centres and scattering centres for electrons. The above is supported by annealing studies. The $In_{0.5}Ga_{0.5}As$ layer is under residual compresive stress in the as-grown samples. Annealing promotes the continuation of the relaxation of the layer, producing more misfit dislocations at the interface and consequently depleting electrons from a larger region [7].

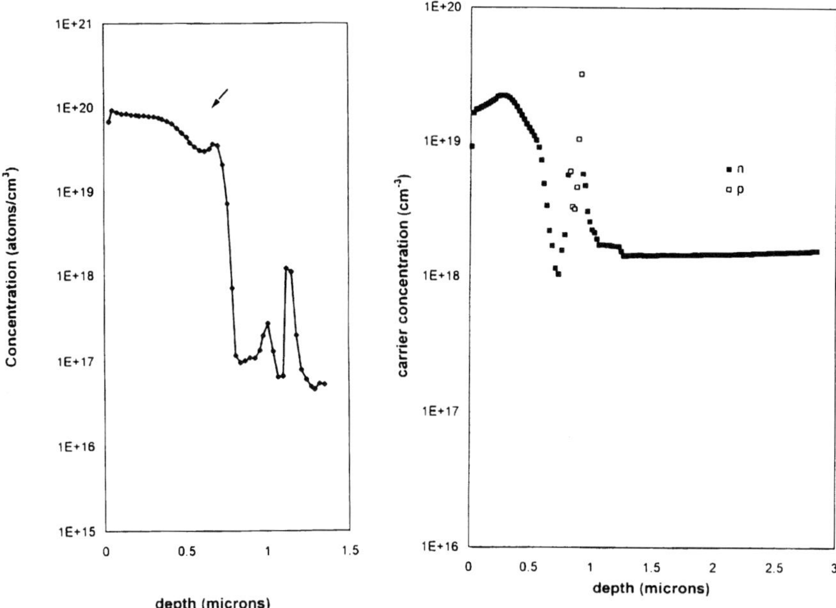

Figure 1: SIMS profile from sample1. An increase in DES flux resulted in increased sulphur concentration

Figure 2: C-V profile from sample 1. Carrier concentration decreases near the InGaAs/GaAs buffer layer interface, whereas a spike-shape increase appears at the GaAs buffer / substrate interface.

Contrary to the decrease of carrier concentration, no interpretation could be proposed for the spike-shaped increases. The most probable explanation is that these spikes are artefacts due to problems in the C-V profiling measurements, arising when the InGaAs / GaAs buffer or the GaAs buffer / GaAs substrate interfaces are etched. The defects present at these interfaces may alter the etch rate as well as other conditions of the C-V profiling process, the effect being more obvious in the case of layers with relatively low doping levels

Microstructural characterisation

Initially, plan-view TEM results confirmed that all the layers studied were partially relaxed, the residual strains in the highly doped samples lying slightly above the empirical curve of Dunstan et al [8]. It was proposed that this reluctance to relax might be attributable to impurity pinning of dislocations [9].

Plan view TEM micrographs of the InGaAs/GaAs interface show the misfit dislocation array along <110> line directions reported by Goodhew [10], and also non-straight dislocation lines and loop-like configurations (figure 3). Such configurations have not been reported previously for sulphur doped InGaAs/GaAs. Dislocation loops and helices, however, have been reported for the case of liquid encapsulation Czochralski (LEC)-grown sulphur doped GaAs and have been attributed to dislocation climb associated with condensation of interstitials [11].

TEM studies were carried out on a range of samples with several values of thickness and sulphur composition. The plan-view TEM micrograph from sample 2 presented in figure 3 indicates that in addition to the misfit dislocation array along the <110> directions, non-straight dislocation lines and loop-like dislocation configurations are formed near the InGaAs/GaAs interface. More dislocation loops and fewer straight dislocation segments were observed in samples with larger sulphur concentration (figure 4), and incipient helical dislocations were observed in the most heavily doped samples. These dislocation configurations are similar to the ones reported by Kim and Lee [12] in the case of Se-doped GaAs.

Kamejima et al [11] studied LEC-grown S-doped GaAs in which they observed dislocation loops of the size of 20-100 nm and helical dislocations as long as several microns. Dislocation climb due to point defects was suggested as the most possible mechanism for the formation of these configurations. Two kinds of point-defect sources were considered: point defects caused by *nonstoichiometry of host material*, and point defects produced so as to compensate for the volume change in the matrix due to *precipitate formation*. Following the model proposed by Abrahams et al [13] for Se precipitation in CVD-grown GaAs, they suggested that initially the diluted ternary alloy $GaAs_{1-x}S_x$ forms, which then decomposes into Ga_2S_3 and nonstoichiometric GaAs matrix crystal:

$$3GaAs_{1-x}S_x \text{ -----> } x\, Ga_2S_3 + 3(1-x)GaAs + x\, Ga$$

Thus, as Ga_2S_3 segregates are formed, point defects such as Ga interstitials are produced.

Plan-view and cross-section STEM studies were carried out in order to provide evidence for sulphur precipitation. The microscope used has one of the highest spatial resolutions in the world for x-ray analysis. Using a probe size of 1nm, STEM should detect precipitates 5nm or larger in diameter. Regions containing helical dislocations or dislocation loops were closely examined for evidence of precipitate formation. No such

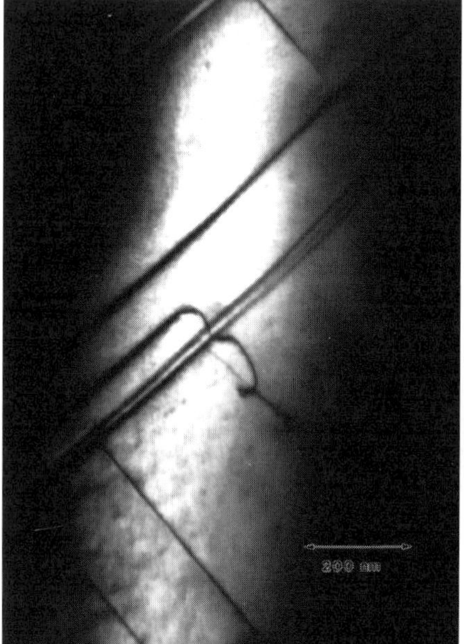

Figure 3: Plan-view TEM micrograph from sample 2 (low doping level). Non-straight dislocations appear alongside the misfit dislocation array.

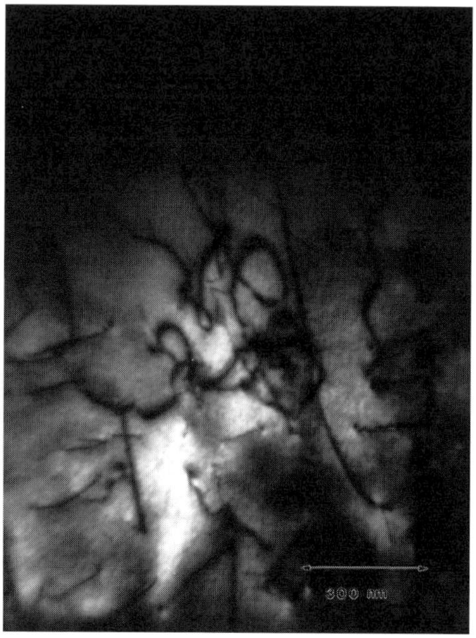

Figure 4: Plan-view TEM micrograph from sample 1. As doping levels increase, more non-straight dislocation configurations are observed.

precipitation was observed, nor was any evidence found for segregation of S to the dislocations themselves. We therefore conclude that point defect generation is not related to the precipitation of sulphur compounds or if they exist the precipitates must be smaller than 5nm in diameter.

Although the present study did not provide evidence for sulphur precipitation, it is the first to report helical dislocations in S-doped InGaAs layers grown by CBE. An explanation of how the dislocation loops and helices form would be of great importance for semiconductor science and technology, as it would result in a better understanding of the interaction between impurities and dislocation behaviour.

CONCLUSIONS

DES was successfully used for S-doping of InGaAs layers. High levels of sulphur incorporation and carrier concentration were achieved, with no evidence of saturation observed. All InGaAs layers studied were partially relaxed, and plan-view TEM micrographs show the misfit dislocation array previously reported in the case of InGaAs/GaAs layers and also non-straight dislocation lines and loop-like configurations, the dislocation behaviour depending on sulphur doping. Helical dislocations were observed by cross-section TEM, indicating that dislocation climb takes place that is associated with condensation of point defects The dislocation loops and helical dislocations observed, although they probably interact with sulphur impurities and affect the relaxation processes (thus the microstructure of the layer), do not appreciably reduce the carrier concentration. Plan-view and cross-section STEM studies did not provide evidence of sulphur precipitation, indicating that if such precipitation occurs, the precipitates are small and dispersed rather than forming large clusters.

ACKNOWLEDGEMENTS

G.M.P. wishes to thank the EU for the provision of a studentship under the Human Capital and Mobility programme.

REFERENCES

[1] KUHN KJ AND DARLING RB: "Mobility of Strained and Dislocated $In_xGa_{1-x}As$ semiconductor material", IEEE Trans. Electron Dev., 1992, **39**, 1288-1294

[2] DOBBELAERE W, DE BOECK J, HERMANS P, MERTENS R, BORGHS G, LUYTEN W AND VAN LAUNDUYT J: "InAs p-n Diodes Grown on GaAs and GaAs-coated Si by Molecular Beam Epitaxy", Appl. Phys. Lett., 1992, **60**, 868-870

[3] HIRTZ JP, "Metal Organic Research in Semiconductor Epitaxy", Mat. Sci. Eng. B - Solid State Materials for Advanced Technology, 1993, **17**, 9-14

[4] JOYCE TB, PFEFFER T, BULLOUGH TJ AND JONES AC, "Metalorganic Sulphur Sources for the Doping of GaAs Grown by Chemical Beam Epitaxy", J. Crys. Growth, 1994, **135**(1-2), 31-35

[5] JOYCE TB, PFEFFER TL, BULLOUGH TJ, PETKOS G, GOODHEW PJ AND JONES AC, "The use of Diethylsulphide for the Doping of GaAs, AlGaAs, InGaAs Grown by Chemical Beam Epitaxy", J. Crys. Growth, 1995, **150**, 644-648

[6]JOYCE TB, *"The growth of GaAs and GaAlAs by chemical beam epitaxy"* PhD thesis, University of Liverpool (1993)

[7] CHANG JSC AND PATTERSON GA, "Effect of Crystalline Defects on Electrical Properties of Heavily Si-doped Strain-relaxed $In_{0.5}Ga_{0.5}As$ Layers Grown by Molecular Beam Epitaxy", J. Vac. Sci. Technol.,1993, **B11** (3), 587-592

[8] DUNSTAN DJ, KIDD P, HOWARD LK AND DIXON RH, "Plastic Relaxation of InGaAs Grown on GaAs", Appl. Phys. Lett., 1991, **59**, 3390-3391

[9] PETKOS GM, GOODHEW PJ AND JOYCE TB, "Sulphur Doping of InGaAs Using Diethylsulphide", J. Crys. Growth, 1996, **164**, 415-419

[10] GOODHEW PJ, "Dislocation Behaviour at Heterointerfaces in III-V Semiconductors", J. Phys. Chem. Solids, 1994, **55**(10), 1107-1114

[11] KAMEJIMA T, MATSUI J, SEKI YASUO AND WATANABE HISAO, "Transmission Electron Microscopy study of microdefects in Dislocation-free GaAs and InP Crystals", J. Appl. Physics, 1979, **50**, 3312-3321

[12] KIM DK AND LEE BT, "Prevention of Substrate Melt-back by Se Addition During Liquid Phase Epitaxial Growth of GaAs on GaAs-coated Si", Materials Letters 1994, **20**, 335-338

[13] ABRAHAMS MS, BLANC J AND BUIOCCHI CJ, "Intedependence of Strain, Precipitation and Dislocation Formation in Epitaxial Se-doped GaAs", J. Appl. Phys., 1974 **45**(8), 3277-3287

THE ROLE OF MISFIT DISLOCATIONS IN THE DEVELOPMENT OF SURFACE STRIATIONS IN STRAINED LAYERS

K. P. GIANNAKOPOULOS* AND P. J. GOODHEW

Materials Science and Engineering, The University of Liverpool, Liverpool L69 3BX, UK.
*Electronic mail: k.giannakopoulos@liv.ac.uk

ABSTRACT

Low-misfit strained epilayers develop a cross-hatch pattern on their surfaces. This is a phenomenon related to the a/2<101> misfit dislocation array, which is introduced during growth on the interface between the epilayer and the substrate. We have studied the surface morphology of III-V epilayers grown on exactly oriented and vicinal plane substrates, and we report for the first time the formation of non-parallel striations for the case of growth on the vicinal substrates. Our layers are $In_xGa_{1-x}As$ with x=0.1 and 0.2 grown by Chemical Beam Epitaxy (CBE) on exactly oriented and on (001) vicinal GaAs substrates, offcut 1°, 2° and 3° towards the [110] and [-110], with thickness varying between 100 and 1000 nm. We have used Nomarski Differential Interference Microscopy (NDIC) and Atomic Force Microscopy (AFM) to study the surfaces and Transmission Electron Microscopy (TEM) to study the dislocation array. Instead of two sets of ridges normal to each other (cross-hatch pattern), we always observe that the set which is parallel to the direction of offcut is divided into two non-parallel subsets that have a small angle between them. We observe this angle whatever the fine structure or the height of the ridges are. This angle is consistent with the theory for the formation of non-parallel 60° misfit dislocation arrays on a vicinal interface. Therefore, we suggest that the formation of the surface crosshatched pattern is directly related to the strain fields around these dislocations and not the edge dislocations, that lie near the interface and exactly parallel to the <110>, as other authors have suggested. Also the possibility of these striations being related to dislocation slip steps has been ruled out.

INTRODUCTION

The heteroepitaxial growth of relaxed buffer layers can have a huge technological importance for modern electronic devices. As many authors have demonstrated[1] it is possible to fabricate lasers grown on such layers. The ideal for these layers would be to have a completely flat surface, but this can be far from true in many cases.

The main reason is that surfaces of low-misfit strained layers that are significantly relaxed (i.e.: more than four times thicker than the Matthews and Blakeslee[2] critical thickness) develop a cross-hatch pattern[3]. It is usually visible with an optical microscope (e.g. with NDIC) or in extreme cases (thick layers) even with the naked eye[4]. If the layer is very thin or has a very low misfit strain an AFM may be required.

It is not yet clear what is the origin of this surface pattern, but it is widely believed that it is related to the glissile misfit dislocations that form the orthogonal dislocation arrays at the interface. The reason for that is not only the similarity of the two patterns but also the fact that the surface cross-hatch is not present on high-misfit layers which relax by edge dislocation formation when the layer is still in the form of islands.

The cross-hatch pattern was first reported by Burmeister[3] in the GaAsP/GaAs heterostructure and since then it has been observed on a plethora of material systems, in both constant and graduated composition layers. It is seen with the use of most growth techniques (e.g. MBE, CBE, LPE, MOVPE), with the only exception of Atomic-Layer Molecular Beam Epitaxy (ALMBE). Nevertheless, even there, it is observed under certain growth conditions[4].

There are many factors that could be considered to be responsible for the origin of the local change in the growth rate that results in a cross-hatched surface. It could be due to impurity concentrations near the dislocations that lie in the layer[5], or more probably due to local strain fields around them. It is well known that strain fields affect the growth rate in epitaxy, producing dots and surface irregularities. Most authors now suggest that the stress concentration is the driving mechanism behind the creation of such surface undulations[6,7]. Theoretical considerations have led to the conclusion that mass transport during growth will occur away from regions of high stress, towards more relaxed regions[8]. Therefore it is useful to look into the details of the MD network inside these layers.

In the low-misfit layers there are usually two types of dislocations which contribute to the strain relief. The most common 60° MD, which forms the MD array in the interface, and the less common edge MD that lies out of the interface. Both dislocations lie parallel to a <110>, with the edge dislocation not lying exactly on the interface but close to it, as it is produced by the glide and reaction of two 60° MDs. If the growth takes place on a vicinal substrate, the 60° dislocations do not lie exactly parallel to the <110> but a small angle away from it, as explained by Kightley et. al.[9]. On this occasion the MD array does not consist of two sets of MDs normal to each other, but the set that lies parallel to the direction of offcut is divided into two subsets that are a small angle either side of the <110>. The angle between them is proportional to the angle of offcut θ. It is: $\vartheta \cong \frac{\sqrt{2}}{2}\theta$.

The average line direction of the edge dislocations in such layers will still be exactly parallel to the <110>. These dislocations are expected to have a greater impact on the formation of the surface morphology than the 60° dislocations because of the larger stress field around them and of the fact that they lie out of the interface, and closer to the surface. Another reason for that assumption is that the approximate spacing of the edge dislocations is usually more than an order of magnitude larger than the spacing of the 60° dislocations, a spacing close to the distance between the surface ridges[10].

Groups (bunches or pile-ups) of 60° MDs might also be responsible for the surface cross-hatch. They are very often mentioned in the literature in order to explain many phenomena such as the CL images (see e.g. Mazzer et. al. 1995[11]) or the blocking of threading dislocations. TEM observations of bunches are rare (for example we can see one given by Albrecht et. al.[12]), but usually a dark line defect in CL is assumed to represent 5-10 dislocations. The similarity between the surface cross-hatch as seen by SEM (or AFM) and the CL maps indicates that there might be a relationship between dislocation groups and the surface ridges. In particular Fitzerald et. al.[13] attribute to groups of dislocations the existence of the surface ridges, the non-radiative lines in the CL maps and the black and white bands of contrast in the plan view TEM images of the interfaces. These bands are also mentioned by Dunstan[14] as the result of bad growth, but to our knowledge they are very common and may arise from nothing more than randomly located sources.

EXPERIMENTAL PROCEDURE

The samples studied were constant composition, single layers of $In_xGa_{1-x}As$ grown on nominally exact and on vicinal (001) GaAs wafers with x=0.1 and 0.2. The vicinal wafers were 1°, 2° and 3° offcut about the [110] and [-110]. All layers were grown by chemical beam epitaxy (CBE) in a VG-V80H CBE system described previously[15]. The precursors used were triethylgallium (TEGa), trimethylindium (TMIn) and arsine (AsH_3) which was cracked to arsenic (As_2) in a high pressure injector cell. GaAs wafers were either mounted directly onto an indium-free sample holder or were indium-bonded on a molybdenum block. There was no pre-treatment and the samples were transferred into the growth chamber through a load-lock chamber. After degassing at 450°C and heat-cleaning of the wafers at 600°C under an As_2 beam, a 250 nm thick GaAs buffer layer was grown at 500°C before the growth of the $In_xGa_{1-x}As$ layer, at a growth rate of 0.28 nm/s at 500°C. The 100, 200, 400 and 1000 nm thick layers were studied using Nomarski Differential Interference Contrast microscopy (NDIC), contact mode AFM (Digital Instruments, Nanoscope II) and TEM (JEOL 2000FX) operating at 200 kV. The TEM samples were prepared by mechanical back thinning to approximately 100μm and thinning to electron transparency by a chlorine in methanol jet. AFM and NDIC samples required no further preparation. TEM cross-sections were prepared in order to confirm the layer thickness.

RESULTS AND DISCUSSION

As the layers grow thicker the surface morphology roughens because of the development of the cross-hatch pattern. The fact that the striations that form the cross-hatch pattern grow higher with increasing layer thickness is well documented[16]. But it has not been reported until now that the cross-hatch pattern on a layer grown on an offcut substrate is different from the one grown on an exactly oriented substrate. The striations on this occasion do not form two sets that are exactly perpendicular to each other, but the set which is parallel to the direction of offcut is divided into two subsets that have a small angle between them (see figure 1). A very interesting observation is that there is not a third set of striations that run exactly parallel to the <110>, in between the two other sets.

We have observed this phenomenon for every layer we have grown, indium bonded or not, whatever the angle or the direction of offcut, and even on layers where the cross-hatch pattern just starts to appear. The angle between the two subsets is always the same as the angle we observe for the two 60° MD subsets that lie on the vicinal interface, parallel to the striations on the surface.

The fine structure of these surfaces is not always the same, but this does not seem to affect the cross-hatch pattern, even when this difference is visible with NDIC. The angle between the striations is visible whatever the fine structure of the surface, just like the striations themselves.

Figure 1: NDIC image of the surface of a 1μm thick $In_{0.2}Ga_{0.8}As$ layer grown on a vicinal (001) GaAs substrate, with an offcut angle θ of 3° ([001] towards the [-110]). The striations along the [110], are exactly parallel to it, where the ones that are perpendicular to them are divided into two subsets with a small angle between them equal to about 4.5°. This angle is almost the same as the angle of the two subsets of the 60° MDs on the interface.

From the above observations we can conclude that it is the original 60° MDs that are primarily responsible for the surface cross-hatch pattern, and not the edge MDs as it has been suggested[17]. If this were the case, then the cross-hatch pattern would always consist of striations exactly parallel or perpendicular to each other, along the <110> just like the edge dislocations. We have to mention here that there cannot be edge dislocations with line dislocation parallel to the 60° dislocations for the simple reason that there cannot be such 60° dislocations that can react in this way, since their reaction involves glide in opposite {111} glide planes. All the parallel 60° dislocations, as we mentioned earlier, have to come from the same glide planes. That proves that the striations observed in our vicinal surfaces are not related to the edge dislocations. In the surfaces of the layers grown on exact substrates we do not see any differences in the density or the height or any other characteristic feature of the striations with respect to the vicinal epilayers. So we can conclude that the striations in both systems are created from the same source, and this source cannot be the edge dislocations. On the other hand, the 60° dislocations seem to play the most important role even at a thickness as high as 10 μm.

The role of the edge dislocations is therefore not so important as originally thought. One reason could be that the edge dislocations are formed at later stage of relaxation when the surface cross-hatch is already present. Another reason might be that it is more difficult for them to affect the growth because of the fact that their average length is relatively small in comparison with the length of the 60° dislocations. In vicinal layers their length is even smaller and it decreases further with the increase of the vicinal angle. On the other hand the increase of the vicinal angle should cause more reactions between 60° dislocations and therefore a larger number of short edge dislocations should exist there.

A first consequence of the observation that there is a difference in the orientation of the striations on a vicinal epilayer, is that now we have an easy way to confirm the angle and the direction of the offcut of the wafer we have grown on, without having to go through time consuming experiments (X-ray or TEM). Laue back reflection radiography is the technique which is routinely used for the detection of the vicinal angle of a wafer and it can now be avoided if we grow a relaxed layer. We can also detect local variations of the vicinal angle in practically the whole surface area of a wafer, without having to cleave it. Basically we can confirm the vicinal angle and orientation in exactly the same way as we could do with the TEM observation of misfit dislocations on the vicinal interface. Furthermore there is no need for the use of a TEM and the sample preparation this requires. The vicinal angle is measurable through the observation of striations which are usually visible with a much cheaper optical microscope.

Another interesting consequence of this observation is that now we can interpret phenomena that we see on a cross-hatched surface (e.g. dislocation blocking) as the side effect of reactions between 60° MDs (or groups of them) as they are reproduced on the surface (see figure 2). This information can be very helpful in understanding the behaviour of the dislocations as we can see evidence of what is happening to the 60° MDs inside the layer, over an area which is practically the whole surface area of the wafer, with the use of only an optical microscope. Also, if the MD density is high, it can be very difficult to resolve dislocation reactions near the interface in TEM plan views, in addition to the requirement for complicated specimen preparation when the layers are very thick (≥1μm). Observation of the cross-hatch

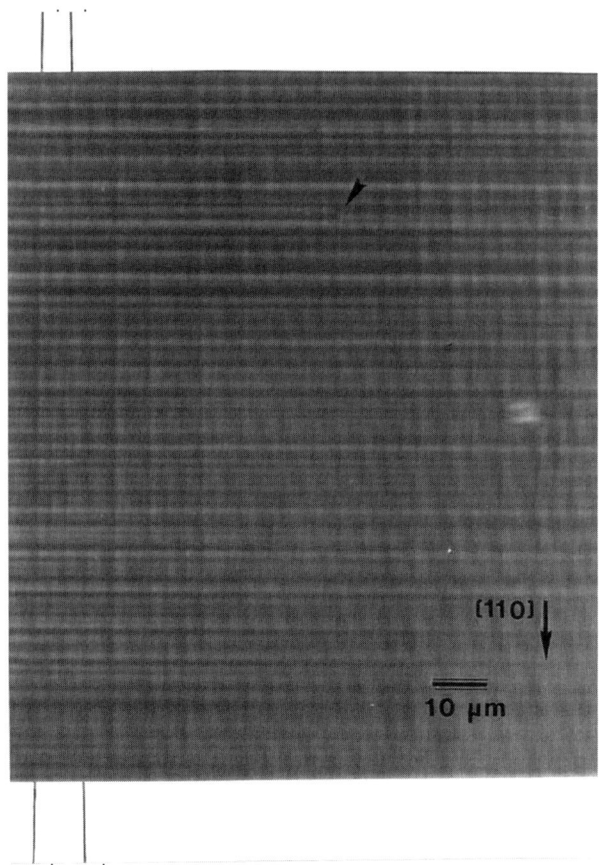

Figure 2: NDIC image of the surface of a 1µm thick $In_{0.2}Ga_{0.8}As$ layer grown on a vicinal (001) GaAs substrate, with an offcut angle θ of 1° ([001] towards [110]). We see again the small angle between the striations on this direction, thought it is not so obvious now. The fact that we are working with very small angles is probably the reason why none has made this observation before. Arrowed is evidence for dislocation blocking.

pattern can make things clearer as it works as a filter for the stronger effects to be seen. It can be the key to the understanding of the dislocation multiplication processes inside the layer[12].

The NDIC images of our surfaces look surprisingly similar to the X-ray Topography (XRT) images of dislocations from Franzosi[18] who among other work studied layers grown on vicinal wafers. This gives strong support to the argument that the cross-hatch pattern gives a picture of the 60° dislocations at the interface. Franzosi comes to the same conclusion with regard to the line direction of the MDs in the interface. XRT is a low resolution technique that can give only a rough idea of the dislocation configuration inside the material, and in this case it is showing groups of 60° MDs in or near the interface.

These groups of 60° MDs could have been generated in the grooves of the cross-hatch pattern as proposed by Albrecht et. al.[7]. But as he mentions, even if they are generated there, they will affect the growth rate at a later stage of the growth process, enhancing the growth on the area above them. We do not know if these groups are just bunched 60° dislocations lying in the interface or pile-ups above it, but as it shown in the next paragraph there is some evidence in TEM observations that it is the latter that is happening.

Figure 3 is a TEM plan view of the interface and the layer whose surface is shown in figure 1. We can clearly see some 60° dislocation segments lying above the interface, within the layer (one of them is arrowed). The fact that they are above the interface is confirmed if we observe the average misfit dislocation density close to them, which is much smaller than the misfit dislocation density in other areas of the sample (below) which were not etched away during the sample preparation process. We believe that these dislocations are on the top of a dislocation pile-up, because if we look along the dislocation line we can see that in the regions where the interface is not completely etched away there are other similar (60° dislocations)

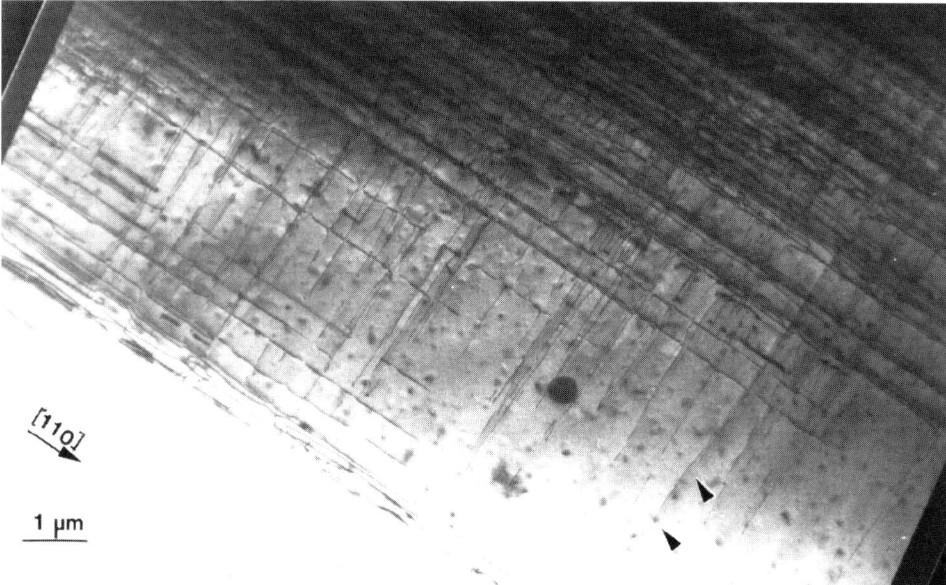

Figure 3: TEM plan view of the sample shown in figure 1. Arrowed is a group of dislocations some of which lie above the interface.

running very close and parallel to it. These dislocations are lying in approximate directions that have an angle of about 4.5° between them, parallel to the striations above. All the dislocations in a pile-up are likely to follow one of the two possible line directions because they probably come from the same source and are identical, and therefore they glide on the same {111} plane.

Additionally, when we observed (see figure 4) one of the 100 nm thick layers using AFM we saw that the cross-hatch pattern has a maximum height of about 10 nm with a spacing that can be as low as ~ 140 nm, approximately the spacing of the 60° MDs on the interface below, as seen in the plan view TEM image. Of course, when the layer grows thicker the striations become significantly higher and have a much larger spacing (about 10 times), making them visible with NDIC. The observation at this early stage indicates that each 60° MD can correspond to one striation as Jonsdottir[19] has predicted. As the layer grows thicker, the distance between the 60° MDs and the surface increases and it can be now a group of 60° dislocations that will affect the growth rate with its larger stress field and create the cross-hatch pattern as we know it for a thicker layer (i.e. with a mean spacing between the ridges of about 1 μm).

Figure 4 reveals that the ridges consist of many irregular atomic steps. The top of the ridges is quite flat and that is why we can see clearly the monoatomic islands on the top. As we see towards the groove between the ridges, they consist of many monoatomic steps that are much more dense. We can not see the steps so clearly towards the groove, because the AFM tip can not penetrate between the islands, as it is deflected from the surrounding features first.

This fine structure of the cross-hatch pattern indicates that even at this early stage the striations are not related to slip steps or enhanced growth around them, such as the ones seen by Wilson et. al.[20] in the $In_{0.5}Ga_{0.5}As/GaAs$ system (7.5 nm thick), Albrecht et. al.[12] (SiGe on Si) or Ressier et. al.[21]. Celii et. al.[22], Park et. al.[23], Howard et. al.[16] and other authors have suggested that the cross-hatch is due to the formation of these atomic steps, but the fine structure, as seen above, shows that probably this is not the case.

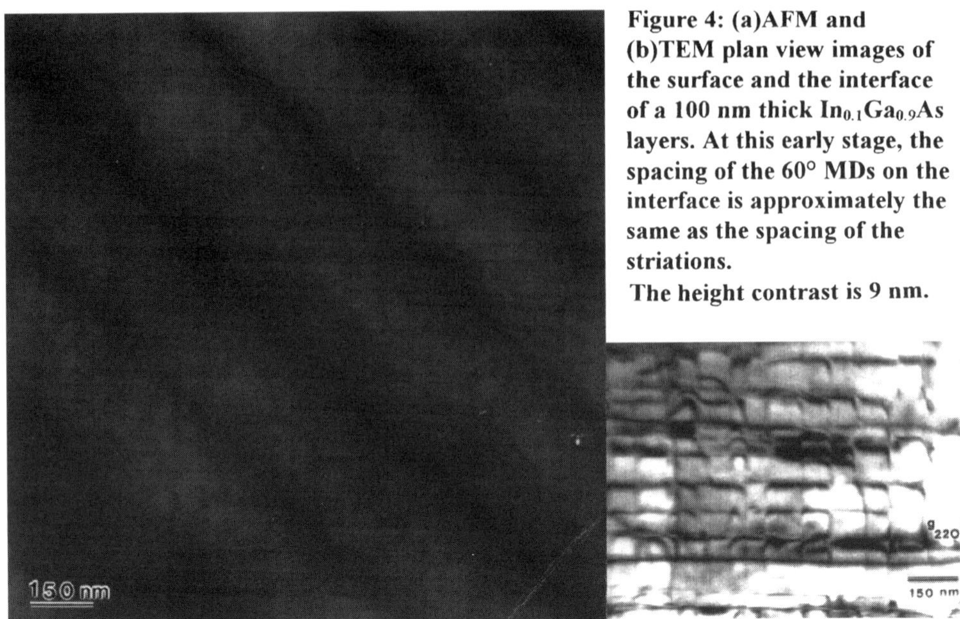

Figure 4: (a)AFM and (b)TEM plan view images of the surface and the interface of a 100 nm thick $In_{0.1}Ga_{0.9}As$ layers. At this early stage, the spacing of the 60° MDs on the interface is approximately the same as the spacing of the striations. The height contrast is 9 nm.

Additionally we can see no reason why dislocation slip steps should affect the growth rate more than the vicinal steps that are much more dense, especially if we make the assumption

that it is single steps (or dislocations) that can cause one striation, as shown in the previous paragraphs. For example, in a layer grown on a substrate which is offcut 1° towards an <110> we expect to see single atomic steps every 16 nm from the first stages of growth, parallel and perpendicular to the slip steps, which at the same stage of growth may have a density of two orders of magnitude less, and are therefore insignificant. We have seen that in such vicinal epilayers the cross-hatch pattern is almost identical to that of the exact layers.

CONCLUSION

We have presented evidence that 60° MDs or groups of them are responsible for the formation cross-hatch pattern and not the edge dislocations that lie in the epilayer. Also the possibility of the surface cross-hatch being related to slip traces has been ruled out with the help of AFM and considering the much higher density of the misfit dislocation array.

ACKNOWLEDGEMENTS

The authors wish to thank Dr T. B. Joyce for the growth of the samples. This work has been carried out at The University of Liverpool as a part of a Community Training Project financed by the European Commission under the Training and Mobility of Researchers (TMR) Programme.

REFERENCES

1. K. MOBARHAN, C. JELEN, E. KOLEV, M. RAZEGHI: 'GaInAsP / InP 1.35 mu-m double-heterostructure laser grown on silicon substrate by metalorganic chemical-vapor-deposition', J. Appl. Phys., 1993, **74** (1), 743.
2. J. W. MATTHEWS, A. E. BLAKESLEE and S. MADER: Thin Solid Films, 1976, **33**, 253.
3. R. A. BURMEISTER, G. PIGHINI and P. GREENE: Trans. TMS-AIME, 1969, **245**, 587.

4. R.BEANLAND, M.AINDOW, T.B.JOYCE, P.KIDD, M.LOURENCO and P.J.GOODHEW: 'A study of surface cross-hatch and misfit dislocation-structure in $In_{0.15}Ga_{0.85}As$ / GaAs grown by chemical beam epitaxy', J. Crystal Growth, 1995, **149**, 1-11.
5. S. KISHINO, M. OGIRIMA and K. KURATA, J. Electrochem. Soc., 1972, **119**, 617.
6. A. G. CULLIS: 'Strain-induced modulations in the surface-morphology of heteroepitaxial layers', Mrs Bulletin, 1996, **21**, 21-26.
7. M.ALBRECHT, S.CHRISTIANSEN, J.MICHLER, W.DORSCH, H.P.STRUNK, P.HANSSON and E.BAUSER: 'Surface ripples, crosshatch pattern, and dislocation formation - cooperating mechanisms in lattice mismatch relaxation', Appl. Phys. Let., 1995, **67**, 1232.
8. C-H. CHIU and H. GAO: 'Mechanisms of Thin Film Evolution', D. J. Yalisove, C. T. Thomson and D. J. Eaglesham ed., Mater. Res. Soc. Symp. Proc., 1994, **317**, 369
9. P. KIGHTLEY, P. J. GOODHEW, R. R. BRADLEY and P. D. AUGUSTUS: 'A mechanism of misfit dislocation reaction for gainas strained layers grown onto off-axis gaas substrates', J. Crystal Growth, 1991, **112**, 359-367.
10. K. L. KAVANAGH, M. A. CAPANO, L. W. HOBBS, J. C. BARBOUR, P. MAREE, W. SCHAFF, J. W. MAYER, D. PETTIT, J. M. WOODALL, J. A. STROSCIO and R. M. FEENSTRA: 'Asymmetries in dislocation densities, surface-morphology, and strain of gainas/gaas single heterolayers', J. Appl Phys., 1988, **64**, 4843-4852.
11. M.MAZZER, F.GHIRALDO, C.ZANOTTIFREGONARA, D.B. HOLT, K. BARNHAM, J. BARNES, R. GREY and J. S. ROBERTS: 'Cathodoluminescence study of misfit dislocations in ingaas/gaas multi-quantum-wells', Inst. Phys. Conf. Ser. **146**, 1995, 741-744.
12. M. ALBRECHT, S. CHRISTIANSEN and H. P. STRUNK: 'The driving-force for dislocation multiplication in the substrate of misfitting heteroepitaxial systems', Physica Status Solidi A Applied Research, 1995, **150**, 453-461.
13. E. A. FITZGERALD, D. G. AST, P.D. KIRCHNER, G.D.PETTIT and J.M.WOODALL: 'Structure and recombination in ingaas/gaas heterostructures', J. Appl. Phys., 1988, **63**, 693.
14. D. J. DUNSTAN: 'Strain and strain relaxation in semiconductors', J. Materials Science Materials In Electronics, 1997, **8**, 337-375.
15. T. B. JOYCE: 'An integrated safety system for CBE', J. Crystal Growth, 1990, **105**, 299.
16. L. K. HOWARD, P. KIDD and R.H.DIXON: 'The effect of growth temperature on plastic relaxation of $In_{0.2}Ga_{0.8}As$ surface-layers on GaAs', J. Crystal Growth, 1992, **125**, 281-290.
17. R.BEANLAND and A.R.BOYD, "The origin of misfit dislocations and the surface cross-hatch pattern in low-misfit strained epitaxial layers", Inst. Phys. Conf. Ser., 1995, **146**, 153.
18 P. FRANZOSI: 'Investigation of crystal defects in iii-v heterostructures by x-ray topography and electron-microscopy', Journal Of Crystal Growth, 1993, **126**, 109-124.
19. F. JONSDOTTIR: 'Computation of equilibrium surface fluctuations in strained epitaxial-films due to interface misfit dislocations', Modelling Simul. Mater. Sc. Eng., 1995, **3**, 503-520
20. I.H. WILSON, J.B.XU and C.C.HSU. 'The epitaxial-growth of compound semiconductors observed by atomic- force microscopy', Inst. Phys. Conf. Ser., 1995, **146**, 649-654.
21. L. RESSIER, F. VOILLOT, M. GOIRAN, J. P. PEYRADE and C. VIEU: 'Fractal analysis of atomic-force microscopy pictures of slip lines on a gaas/gaalas heterostructure plastically deformed to obtain quantum wires', J. Appl. Phys., 1996, **79**, 8298-8303.
22. F.G.CELII, L.A.FILES-SESLER, E.A.BEAM and H.-Y.LIU: 'In-situ detection of InGaAs strained-layer relaxation by laser-light scattering', J. Vac. Sci. Technol. A, 1993, **11**, 1796
23. S.J.PARK, J.R.RO, J.S.HA, S. B. KIM, H. H. PARK, E. H. LEE, J. Y. YI and J. Y. LEE: 'Surface-morphology of ingaas on gaas(100) by chemical beam epitaxy using unprecracked monoethylarsine, triethylgallium and trimethylindium', Surface Sci., 1996, **350**, 221-228.

LOCAL ELECTRONIC STRUCTURE OF DEFECTS IN GaN FROM SPATIALLY RESOLVED ELECTRON ENERGY-LOSS SPECTROSCOPY

M. K. H. NATUSCH, G. A. BOTTON and C. J. HUMPHREYS
Department of Materials Science and Metallurgy, University of Cambridge, Pembroke Street, Cambridge, CB2 3QZ, U.K.

ABSTRACT

The optical properties and their modification by crystal defects of wurtzite GaN are investigated using spatially resolved electron energy-loss spectroscopy (EELS) in a dedicated ultra-high vacuum field emission gun scanning transmission electron microscope. The calculated density of states of the bulk crystal reproduces well the features of the measured spectra. The profound effect of a prismatic stacking fault on the local electronic structure is shown by the spatial variation of the optical properties derived from low-loss spectra. It is found that a defect state at the fault appears to bind 1.5 electrons per atom.

INTRODUCTION

Recently the interest in the development of GaN based high-efficiency optoelectronic devices for blue light emission has risen greatly. One surprising and as yet unexplained observation is that GaN can be highly defective and still emit light. Defects and surfaces in semiconductors can introduce defect states in the forbidden energy gap and act as centres for electron-hole pair recombination. The fact that GaN emits light strongly even though it has a high defect density suggests that either the defects do not give rise to states in the band gap, or that they do have states in the band gap but the cross-section for recombination is low, in which case the light emission would be even brighter without these defects. It is therefore important to understand the effect of defects on the local electronic structure of GaN.

Electron energy-loss spectroscopy (EELS) with a field-emission gun scanning transmission electron microscope (FEG-STEM) has the potential to provide information on the modification of the electronic structure by defects at high spatial resolution on a nanometre scale. Hence defect states of individual defects can be examined either at low energy losses by extracting dielectric information from the loss function via Kramers-Kronig transformation or at high losses via the fine structure of inner shell electron excitation edges, for instance of the Nitrogen 1*s* edge. Whereas the former approach provides the joint density of states in a momentum-resolved measurement, the latter shows the angular projected density of states. The experimental results can be compared directly with *ab initio* band theory calculations.

We have developed the methodology to scan the electron beam of the microscope across points of interest in the sample while at the same time acquiring spectra. These spectra are then automatically transformed via Kramers-Kronig analysis to show the real and imaginary parts of the dielectric function of the material under investigation. The results from this operation allow one to calculate optical properties of interest such as the reflectivity as a function of energy and position on the sample. Sum rules also enable the calculation of the effective number of valence electrons, again as a function of energy and position. Thus a complete picture can be drawn of the macroscopic properties of GaN as a function of its microstructure and the method also provides deep insight into the conduction mechanism of the material.

In the following we discuss in greater detail the spatial resolution of EELS and show that spectroscopy of individual defects is indeed possible. We then present low loss spectra from scans across a grain boundary and show how the spectra and the derived optical properties change at the boundary with respect to the bulk. Calculations for bulk GaN inner shell excitatations are presented and compared with experiment. We conclude with a caveat for optical spectroscopy by comparing experimental results at high and low spatial resolution.

SPATIAL RESOLUTION IN EELS ON STEM

The measurements in this paper were performed on a VG STEM HB501, a dedicated STEM with a cold field emission gun, operated at 100 kV. Several instrumental parameters determine the success of a spatially resolved EELS experiment on this microscope. These are most importantly the spatial resolution attainable and the energy resolution of the electron gun, the spectrometer and the detector. The attainable energy resolution is mainly limited by the gun and is close to 0.4 eV over the time of a typical experiment.

The most obvious limit to the spatial resolution in an EELS experiment derives from the limited probe size attainable in a particular electron optical configuration. The competing effects of spherical aberration on one hand and diffraction and the initial virtual source size on the other must be balanced by finding those conditions which provide the optimum convergence angle to obtain the minimum probe size. The convergence angle is a function of the virtual objective aperture and the condenser excitation. The geometric broadening of the beam in the specimen in EELS is determined by the solid angle subtended by the collector aperture and is linear with specimen thickness. The resulting limit to the spatial resolution is then their product. In addition to theses effects, inelastic scattering is delocalised by a distance defined by the impact parameter. Even though more detailed calculations are available [1], a simple approximation [2] is sufficient for the present purpose. Fig. 1 shows the delocalisation expected in the present experiment. The electron beam was scanned across the edge of a GaN specimen into vacuum and energy-loss spectra were acquired during this process at regular intervals. The zero-loss peak was removed from the spectra and the energy-loss at which the inelastic intensity disappears for a given distance recorded. The value of the distance was adjusted for the probe size measured under the same experimental conditions by observing the change in the annular dark field signal while scanning across the edge of a carefully oriented MgO smoke cube.

Assuming the contributions to the spatial resolution are gaussian it is possible to add the electron optical resolution, the aperture function and the impact parameter in quadrature. The resulting spatial resolution in the experimental configuration used here as a function of the STEM C_1 condenser current is given in Fig. 2 for two energy losses, in the low-loss region at the GaN band gap of 3.4 eV and for higher losses at the nitrogen K-edge at about 400 eV. The spatial resolution at low energy losses can be improved by collecting the EELS signal at a high momentum transfer [1]. This approach is not useful here since we are only interested in vertical interband transitions.

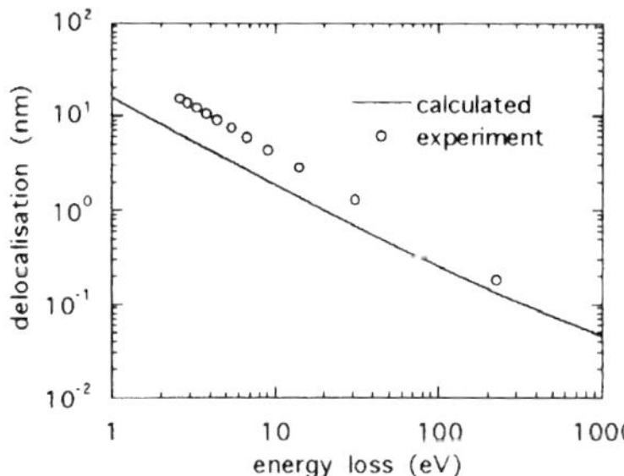

Fig. 1 Delocalisation of inelastic scattering measured in GaN and calculated from [1].

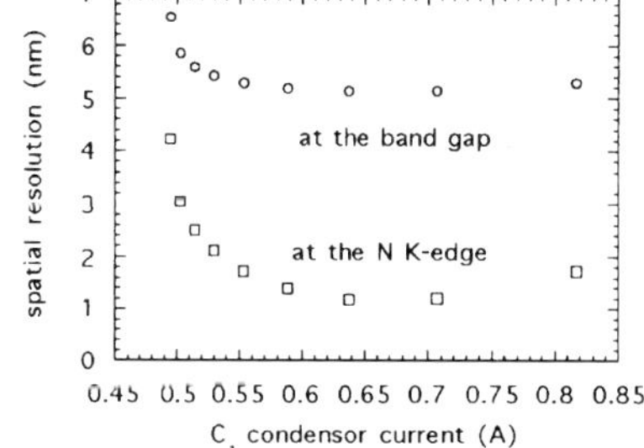

Fig. 2 Ultimate spatial resolution attainable in present STEM at 3.4 eV and 400 eV energy-loss under given experimental conditions.

LOW LOSS SPECTROSCOPY

The low loss spectrum arises from the excitation of valence band electrons to empty states in the conduction band and from collective excitations. The joint density of states between the valence and conduction band is weighted by the matrix element which makes a direct interpretation of a low loss spectrum difficult. Peaks in the joint density of states occur where bands for the initial and final states in the bandstructure are parallel. In terms of the dielectric response function ε with real part ε_1 and imaginary part ε_2, EELS probes the loss function $\mathrm{Im}(-1/\varepsilon)$ which gives a complete description of the dielectric response of the medium to the probe electron. ε_1 and ε_2 combine as $\varepsilon_2/(\varepsilon_1^2+\varepsilon_2^2)$ to form the loss function. Unlike in the optical case, ε is different from the relative permittivity of the medium because of the considerable momentum transfer in electron scattering. Only spectra which are acquired using a small collection angle, so as to accept only a small range of momentum transfers around zero transfer can be directly compared with optical data. The real and imaginary parts of the dielectric function can be extracted from a single scattering distribution by using the Kramers-Kronig relations.

Since a Fourier method is employed for the Kramers-Kronig transformation [3], the high energy end of the measured energy-loss spectrum needs to be extrapolated to zero using an inverse power law to ensure well-behaved Fourier-transforms. A single scattering distribution is retrieved from the experimental spectrum with the Fourier-log deconvolution method [3] which also removes the field emission zero-loss peak. The Kramers-Kronig analysis is iterated several times to remove the effect of surface excitations and the self-consistency of the data is verified by Kramers-Kronig sum rules.

The wurtzite GaN epilayer was grown on $\{\bar{1}\bar{1}\bar{1}\}$B GaP by molecular beam epitaxy. A plan view sample was prepared by ion-milling from the substrate side. Further characterisation of the sample and its grain boundaries has been published previously [4]. After plasma cleaning for 6 minutes, the specimen was immediately inserted into the VG STEM HB501 and heated over a lamp for 1 hour in high vacuum in order to avoid the effects of hydrocarbon contamination. The specimen was then transferred into the column and pumped down overnight to about 10^{-9} torr. There were no signs of contamination or radiation damage visible during the experiment. The maximum accepted momentum transfer was about 17 nm^{-1}. Spectra were acquired at a dispersion of 0.1 eV per channel from a 70 nm thick region of the sample for 50 ms using a UHV compatible Gatan Imaging Filter. The scanning of the electron beam was controlled by Gatan DigiScan which also allows the resulting energy-loss spectra to be transferred to a Gatan Digital Micrograph file for processing. The optical axis of the specimen was perpendicular to the substrate plane and the anisotropic nature of the dielectric function was therefore not visible.

Fitting the region between 7 eV and 13 eV of the loss function to the expected approximately parabolic shape of the density of states curve yields the band gap as 3.4 ± 0.3 eV. The volume plasmon in wurtzite GaN can be calculated to be at 22.0 eV assuming no Ga 3*d* electrons are in the valence band. The presence of the band gap shifts this to 22.3 eV. ε_1 crosses zero at 18.5 eV as can be seen in Fig. 3. The onset of the Ga 3*d* excitation has been reported to be 19.7 eV [5]. It therefore seems that the lower one of the two peaks at 19.4 eV and 22.0 eV is the volume plasmon shifted by the Ga 3*d* and interband transitions which also cause this unexpected double peak. No clear estimate of the band gap can therefore be made from the position of the plasmon. The optical properties as given in Fig. 3 agree well with published measurements and calculations [6, 7].

Of particular interest is Fig. 3(d), the effective number of valence electrons per atom on and off the boundary. The difference between bulk and boundary quickly rises to and remains almost constant at about 1.5 electrons. This indicates that a defect state in the grain boundary binds about 1.5 electrons. The resulting screening from this build-up of charge is visible in Fig. 4 which shows a contour plot of the imaginary part of the dielectric function as a function of distance from the grain boundary and the energy-loss. The lines on either side of the line on the boundary are strikingly different from the remainder. This can be expected because of the conservation of charge neutrality.

We are confident that the defect state arises from the crystal structure of the defect and not from impurities trapped at it. SIMS profiles [8] have shown that impurities are only present below background level. Oxygen and phosphorous are only present below 10^{19} atoms per cm^3, crbon below 10^{18} atoms per cm^3 and silicon and sulphour only below 10^{17} atoms per cm^3. We have unsuccessfully attempted to detect any possible segregation effects of these impurities.

An explanation for the formation of this defect state must therefore be sought from the crystal structure of the defect. Gallium vacancies for instance could well charge the defect with 1.5 electrons per atom. We intend to compare our results with minimum energy calculations for the crystal structure of the prismatic stacking fault [9] to find the cause for the charge on it.

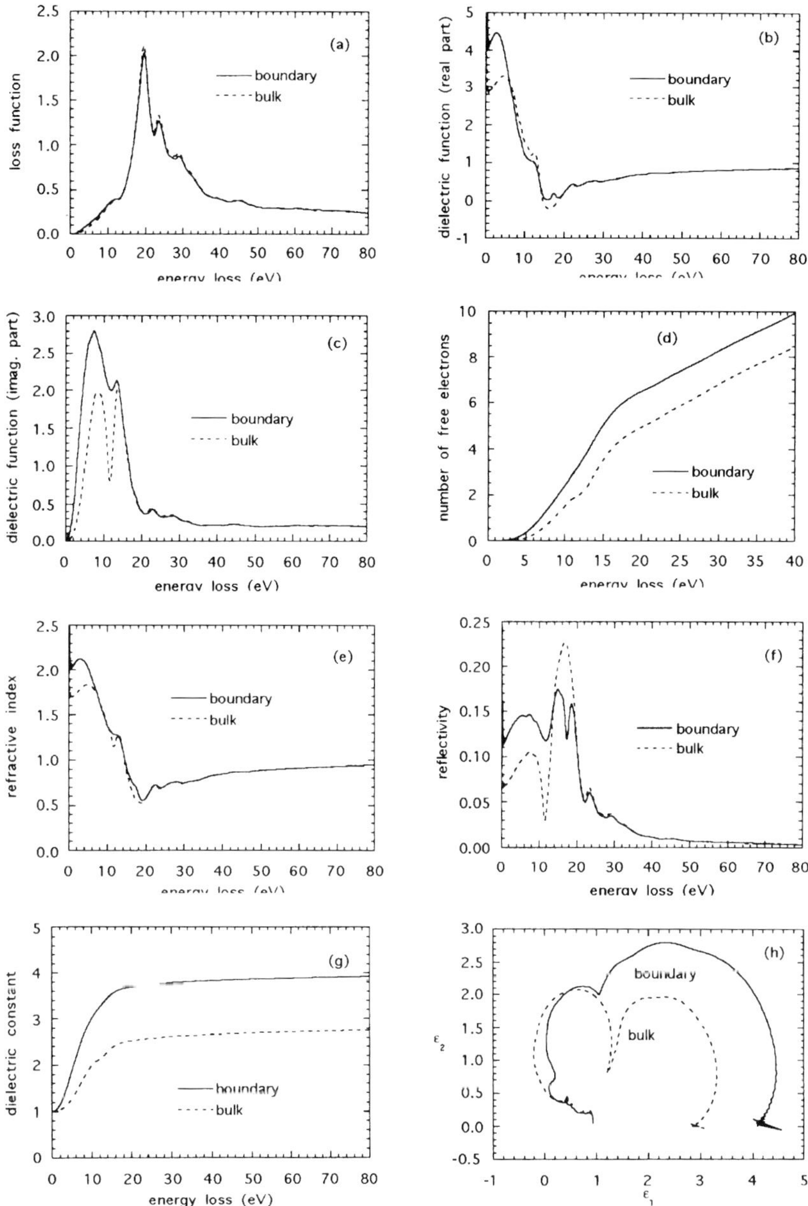

Fig. 3 Experimental results in the low loss region. Shown are the optical properties of GaN as derived from energy-loss spectra by Kramers-Kronig analysis. Only the results on the prismatic stacking fault and far from the fault are shown. (a) loss function (b) real part of the dielectric function (c) imaginary part of the dielectric function (d) effective number of valence electrons per atom (e) refractive index (f) reflectivity (g) effective dielectric constant (h) imaginary part of the dielectric function versus real part of the dielectric function (Cole-Cole plot).

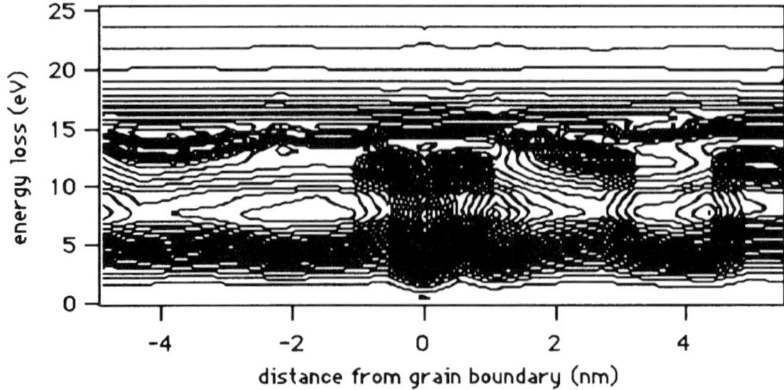

Fig. 4 Contour plot of the imaginary part of the dielectric function as a function of distance from the boundary.

A state in the band gap would be visible effectively as a reduction of the gap. To measure this effect, we have fitted a theoretical shape of the loss function to experimental spectra. The local joint density of states is proportional to $(E - E_g)^{1/2}$, where E is the energy-loss and E_g is the band gap. A theoretical model of the zero-loss peak or an experimental peak has been deconvoluted from the spectrum. Once this contribution is removed, the band gap could in principle be obtained from the zero of the loss function. A further complication is the occurrence of radiation losses [10] and the presence of a large number of interband transitions [6] as can bee seen from Fig. 5.

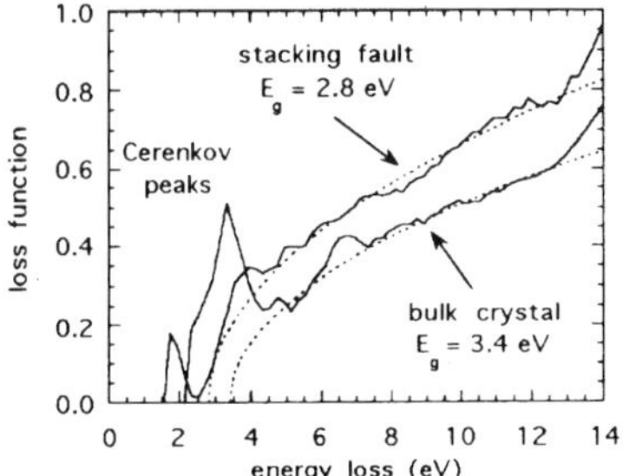

Fig. 5 Low-loss spectra on and off the defect, full lines are the experiment, dotted lines the parabolic fit.

Fig. 5 shows raw spectra acquired from the bulk crystal and the stacking fault with the zero-loss peak removed. The sample was about 30 nm thick in this region and the spectrum was acquired for 32 ms at 0.02 eV per channel dispersion. The spectra have been scaled to the respective loss functions. To obtain statistically meaningful values for the band gap, spectra have been acquired at various different regions in the bulk and on the defect. Specimen drift has been corrected for and the spectrometer has carefully been kept in focus. No contamination effects were visible.

The numerical fitting was performed in the region between 7 eV and 13 eV and yields a band gap of 3.4 ± 0.3 eV in the bulk and 2.8 ± 0.2 eV on the defect. The limited accuracy in this measurement can be explained not only by the arguments given above but perhaps also by the presence of starin in the sample. The change in the band gap is likely to derive from a change in the bandstructure.

CORE-LOSS SPECTROSCOPY

For core losses the initial state of the excited electron is given by an atomic inner shell wavefunction. The final state is given by the wavefunction for an empty state in the conduction band. In the limited region of the edge onset over which the fine structure is observed and calculated, the transition matrix element varies only slowly so that most fine structure arises from changes in the density of states above the Fermi level. Since the cross-section of an inner shell excitation is considerably smaller than that of a valence electron excitation, ε_2 becomes small at high losses. ε_1 tends to unity and the loss function and hence the fine structure therefore become proportional to ε_2.

The density of states of bulk GaN is calculated using the *ab initio* self-consistent linear muffin-tin orbital method within the local density approximation. Using the unoccupied part of the density of states and the relevant transition matrix elements, energy-loss spectra can be calculated. A comparison of these with experimental energy-loss spectra from bulk material shows good agreement as can be seen in Fig. 6.

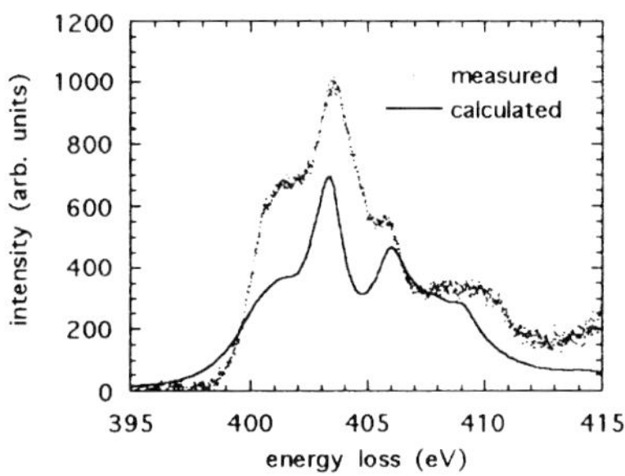

Fig. 6 Calculated and experimental N K-edge.

Band theory assumes an infinite perfect crystal lattice and is therefore not applicable for crystal defects (unless a supercell method is used). However, peaks in the angular projected density of states arise from flat energy bands so that variations of these bands by a defect should be detectable experimentally as a deviation from the calculated edge. Line scans of the N 1*s* excitation acquired from the bulk across the stacking fault revealed an additional peak in the spectrum on the defect.

The distance of the additional peak in the N 1*s* excitation from the original onset of the edge is 1.1 eV ± 0.2 eV. This suggestes the presence of a state in the band gap with this distance from the conduction band or from the valence band. Using low-loss spectroscopy we have obtained a shift in the band gap of 0.6 eV ± 0.3 eV. These two values should agree and one can see that they do so within one standard deviation. The position of this state in the band gap can be compared with the expected result from the minimum energy calculation [9].

PLASMA FREQUENCY

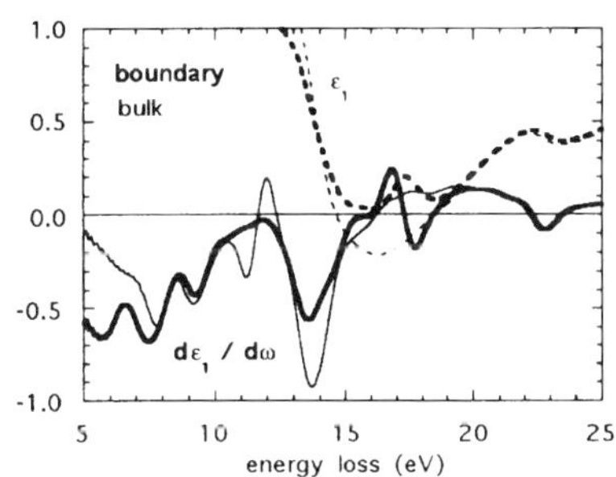

Fig. 7 Real parts of the dielectric functions and their derivatives w.r.t. energy on and off the boundary.

Interestingly the plasma frequency appears to change only slightly as can be seen from Fig. 7 which shows the real part of the dielectric function and its derivative with respect to energy on and off the prismatic stacking fault. This change is expected since, as Fig. 3(d) indicates, an electron is more tightly bound at the defect. It is therefore more localised and because of the uncertainty principle occupies a larger region in moentum space. The bandstructure is less curved and the effective mass, another factor in the plasma frequency besides the number of valence electrons, increases. It is, however, questionable to which extent one can assign a plasmon to a defect.

EELS AND OPTICAL SPECTROSCOPY

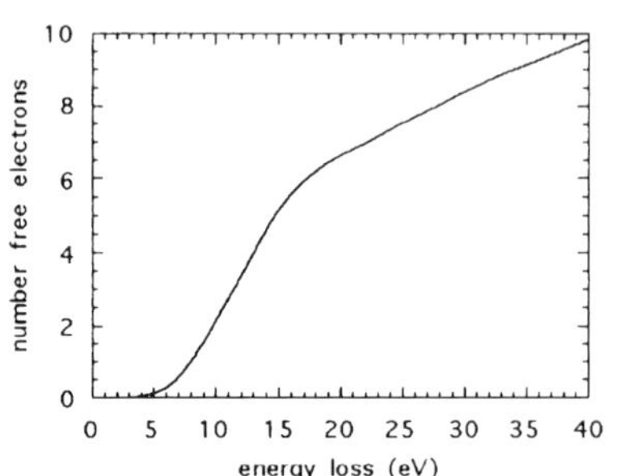

Fig. 8 Effective number of valence electrons acquired at low spatial resolution.

As shown above, EELS provides high spatial resolution and can therefore distinguish between bulk and defect scattering. In an EELS experiment with low spatial resolution, e.g. about 1 μm, the resulting signal can be expected to be a mixture of scattering from both bulk and defects. The effective number of valence electrons resulting from such an experiment is shown in Fig. 8 and is identical to results from synchrotron ellipsometry [7]. This suggests that the low spatial resolution of conventional optical spectroscopy implies that bulk properties alone cannot be investigated. This might help to explain some of the published differences between theory and experiment.

CONCLUSIONS

We show that EELS is a useful tool for the study of the local electronic structure of materials. Whereas optical spectroscopy suffers from low spatial resolution and hence experimental results always contain a mixture of bulk and defect scattering, EELS can separate the two. The EELS experiment reveals the profound effect grain boundaries have on the electronic structure of GaN by binding about 1.5 electrons per atom to a defect state. This makes the intense emission of intense blue light from GaN LEDs even more surprising.

ACKNOWLEDGEMENTS

The authors wish to thank Y Xin for the provision and preparation of the GaN specimen which has been grown by CT Foxon and TS Cheng at the University of Nottingham. MKHN gratefully acknowledges financial support by Gatan Inc. and the EPSRC.

REFERENCES

1. D. A. Muller and J. Silcox: 'Delocalization in inelastic scattering' *Ultramicroscopy* **59**, p. 195 (1995)
2. S. J. Pennycook: 'High resolution electron microscopy and analysis' *Contemp Phys* **23**, p. 371 (1982)
3. R. F. Egerton: 'Electron energy-loss spectroscopy in the electron microscope', 2nd ed., Plenum Press, New York, 1996
4. Y. Xin, P. D. Brown, C. J. Humphreys, T. S. Cheng and C. T. Foxon: 'Domain boundaries in epitaxial wurtzite GaN' *Appl Phys Lett* **70**, p. 1308 (1997)
5. J. Hedman and N. Märtensson: "Gallium nitride studied by electron spectroscopy' *Phys Scr* **22**, p. 176 (1980)
6. S. Bloom, G. Harbeke, E. Meier and I. B. Ortenburger: 'Band structure and reflectivity of GaN' *Phys Stat Sol B* **66**, p. 161 (1974)
7. S. Logothetidis, J. Petalas, M. Cardona and T. D. Moustakas: 'Optical properties and temperature dependence of the interband transitions of cubic and hexagonal GaN' *Phys Rev B* **50**, p. 18017 (1994)
8. S. V. Novikov, T. S. Cheng, C. T. Foxon: personal communication (1998)
9. J. Elsner, R. Jones, M. I. Heggie: personal communication (1998)
10. C. Festenberg: 'Energieverlustmessungen an III-V-Verbindungen' *Z Phys* **227**, p. 453 (1969)

LIQUID PHASE EPITAXIAL GROWTH OF $Cd_xHg_{1-x}Te$ FOR INFRARED DETECTION

P. CAPPER, JOANNE GOWER, C. MAXEY, E. O'KEEFE*, J. HARRIS, L. BARTLETT, S. DEAN
e-mail pete.capper@gecm.com
GEC-Marconi Infra-Red Ltd., P.O. Box 217, Millbrook, Southampton, Hants. SO15 0EG, UK
* Present address: DERA, Farnborough, Hants

ABSTRACT

The current status of liquid phase epitaxy, LPE, of the infrared (IR) detecting material cadmium mercury telluride, $Cd_xHg_{1-x}Te$ (CMT) is described. The basic process is centred on a tellurium-rich melt and the use of high quality CdZnTe substrates, both 20×30 and 30×40 mm in size. Growth takes place in a high purity graphite boat in flowing Pd-diffused hydrogen. A separate HgTe source provides the necessary Hg overpressure control. Growth occurs by ramp cooling from ~500 °C to produce layers with thicknesses between 25 and 35 mm. An in-situ anneal at ~300°C following growth is used to set the acceptor concentration to the level required for photodiode fabrication. Assessment of the layers includes Fourier transform infrared (FTIR) spectrometry to determine cut-on wavelength (related to x) and layer thickness. FTIR spectrometry is also used to map the lateral variations in wavelength and thickness. There is a grade in x through the thickness of the layer and this must be determined and allowed for during device processing. Defect densities should also be kept to a minimum to reduce the numbers of poor or dead diodes on large area focal plane arrays. A further measure of the quality of the LPE layers is the background donor level. This is determined by Hall effect measurements after a low temperature Hg anneal. Levels of $< 1\times10^{15}$ cm^{-3} are normally required for photodiode arrays, while lower levels ($< 4\times10^{14}$ cm^{-3}) are necessary to make the highest performance photoconductive devices. Various chemical analysis techniques have been developed to assess the purity of the material grown and an example of an impurity survey by mass spectrometry will be presented.

INTRODUCTION

The IR market has opportunities for both long linear and two dimensional infrared detectors with peak spectral sensitivities ranging from 2 to 12.5 µm [1, ch. 15]. GMIRL has developed a flexible focal plane array (FPA) technology to satisfy this market, including material and photodiode technologies, read-out circuitry, hybridisation methods and cryogenic encapsulation.

Epitaxial growth requirements include the mechanical dimensions and tolerances of the film, usable area, wavelength uniformity, defect density, through-thickness composition variations, acceptor level (Hg vacancies) and background impurities. Current FPA requirements have increased the size of single pieces of device quality CMT up to 17.5×2 mm for long linear arrays and 6×8 mm for 2D. Until recently, reliable supplies of substrates have been limited to 2×3 cm. However, 3×4 cm substrates are now available commercially and the growth process has been scaled up to take advantage of this.

PROCESS OUTLINE

Epitaxial films of CMT are grown onto CdTe, CdZnTe or CdTeSe substrates from small melts of Cd and Hg dissolved in Te using a horizontal slider technique [1, ch. 2]. Figure 1 shows the in-house designed LPE boat which is made from purified, high density, small grain, electronic grade graphite. The long component is the moving section, or `slider', and it has a precisely machined well in which the prepared substrate is placed. Resting directly on this section is the melt holder which has two large through-wells, for growth and annealing, and a third blind well between. The pre-compounded growth solution is placed in the growth well, and crushed HgTe is placed in the blind well. The HgTe acts as a buffering source of the volatile elements to help stabilise the growth solution composition and also to deliver an overpressure to the annealing well to control the metal vacancy level during annealing. All three wells are connected with a gas channel plate which controls the movement of the volatile elements within the boat. The growth solution and HgTe overpressure source are made from high purity elements; the tellurium is zone-refined in-house in hydrogen and the mercury is distilled in-house, just prior to use. The control over the impurity levels in the major constituent elements is a crucial part of the control of the overall process.

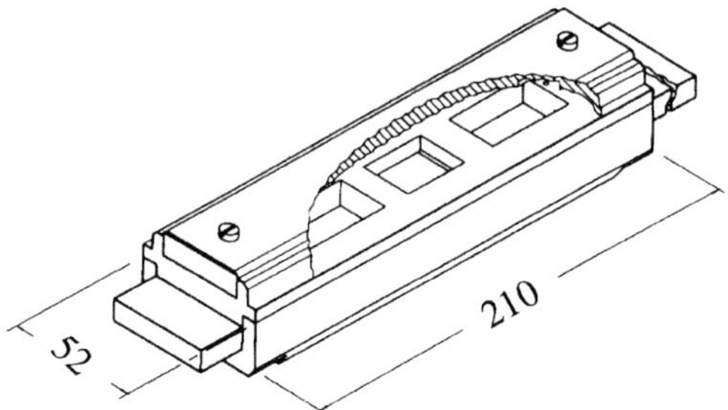

Figure 1. LPE boat design, dimensions in mm.

At the start of the growth cycle the slider is positioned so that the substrate is under the annealing well. The loaded boat is placed in the reactor tube which is then sealed and purged with high purity nitrogen and then Pd-diffused hydrogen. The furnace is pre-heated to 520°C and then moved over the boat. The boat heats rapidly, Figure 2, and after a solution melting and equilibration period, the furnace is cooled rapidly by ~20°C, then a slow cooling ramp (2 or 3°C h^{-1}) is initiated. When the boat reaches the required growth start temperature, the slider is moved so that the substrate is positioned under the molten solution. The ramp continues until the required film thickness has been grown, typically over a 10°C temperature drop, after which the slider is returned to the starting position and the furnace temperature reduced rapidly to the annealing temperature. Following the anneal, the furnace is moved back to its starting position and the system is allowed to cool.

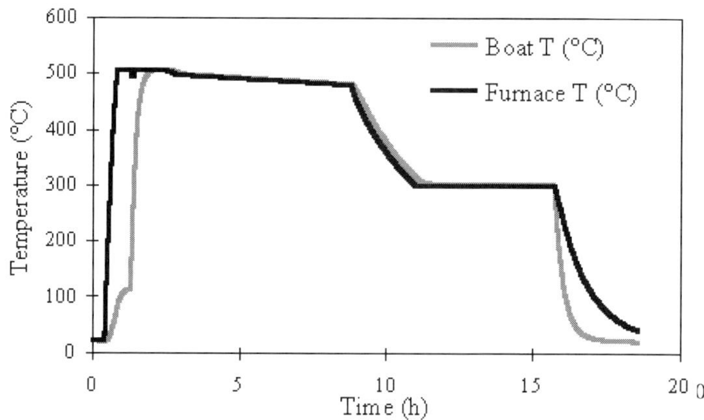

Figure 2. Time-temperature plot for a typical LPE growth run.

COMPOSITION AND THICKNESS CONTROL

The cut-on wavelength of the material is controlled by its composition, expressed as the mole fraction x of CdTe in $Cd_xHg_{1-x}Te$. The composition of the layer is, in turn, controlled by the composition of the melt close to the growth interface, and the segregation coefficients of the elements across the growth interface [1, ch. 2].

A relationship between the FPA spectral response and the starting composition of the growth solution, $((Cd_zHg_{1-z})_{1-y}Te_y)$, has been established over the wavelength range 2.0–12.5 μm at 77 K, for x = 0.45 to x = 0.20, and z = 0.12 to z = 0.42. Variations in the thermal environment and starting solution stoichiometry control the resulting composition which needs to be typically x = 0.21±0.004 for long wavelength material. Temperature during growth needs to be controlled to within 0.2°C. Figure 3 shows an example of a wavelength map (generated using transmission measurements from an FTIR spectrometer) of a 20×30 mm layer demonstrating the high degree of uniformity possible. The transmission measurements also give the thickness of the layer at each point.

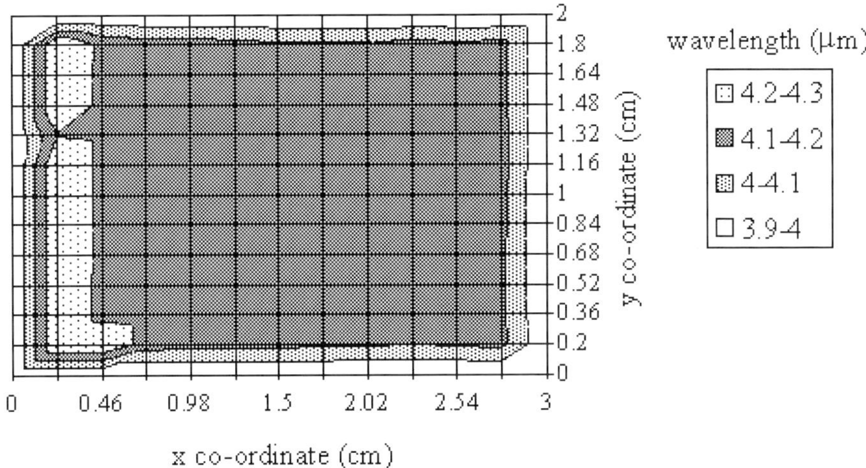

Figure 3. Wavelength variation across a 20×30 mm LPE layer.

SURFACE MORPHOLOGY

Surface morphology is controlled at two levels: micro-texture and long-range variation. The micro-texture is a result of mis-alignment of the substrate crystal plane with the growing surface. Deviations > 0.1° from the <111>B plane lead to significant surface texture. Growth on accurately orientated substrates gives a specular surface on which atomic-scale growth features can be seen using atomic force microscopy (AFM). Figure 4 shows a classical Frank-Read site on an as-grown LPE layer surface.

Long range morphology is controlled by edge effects, substrate flatness prior to growth and control of the growth rate. A 30 μm thick layer usually has a 1 mm wide edge zone of enhanced growth, all of which is removed during device processing. Flatness measurements, using a scanning laser system, reveal typical variations of ~ 5 μm over a 2×3 cm LPE layer, with the majority of this confined to the corners.

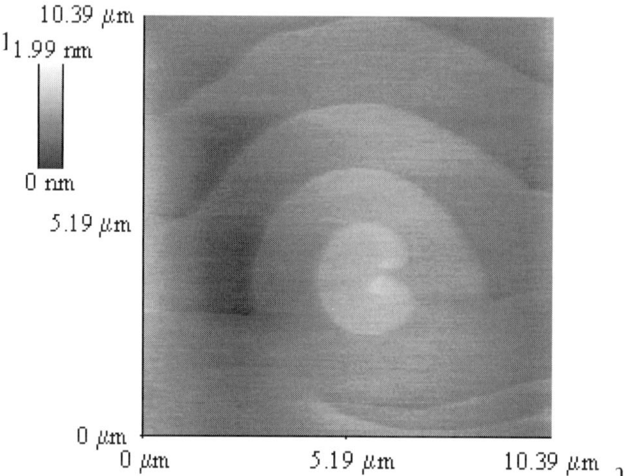

Figure 4. AFM image of a Frank-Read growth spiral.

DEFECT DENSITY CONTROL AND COMPOSITION THROUGH THE LAYER

Defects act as shunt resistors across diode junctions, so dislocations extending into the active CMT need to be minimised. Defect densities above 1×10^5 cm^{-2} can produce diode degradation and this is only slightly above the defect level in the substrates. Figure 5 shows a composition depth profile of a layer measured using Auger electron spectroscopy (AES) on a chemically bevelled sample [2]. Linear AES measurements at 20–100 µm intervals were taken along the bevels, following a light ion beam etch in the AES to clean the sample surface. The Cd/Te signal ratios were calibrated against bulk samples with known x values to give an accuracy of ~ 3% (±0.006 at x = 0.2). The depth resolution was 0.06 µm. There are four distinct sections of the curve:

- the substrate ($t < -1$ µm),
- the inter-diffused region -1 µm $< t <$ 1 µm,
- the graded region 1 µm $< t <$ 7 µm
- the uniform composition region 7 µm $< t <$ 28 µm (Figure 6a).

If the chemically bevelled sample is defect etched [3], three distinct bands of defect structure are seen in the CMT. In the inter-diffused region the etch pit density (EPD) is very high (Figure 6b) and in the graded region the individual etch pits re-distribute into 3-fold symmetric patterns on the <111> planes (Figure 6c), whereas the region of the layer used for devices has a very low EPD, (Figure 6d). The graded region is thought to modify the strain profile and this effect promotes dislocation climb and annihilation. In layers where the graded region is confined to 7 µm of the interface, clearly defined cross-hatching and low final EPD ($1–5\times10^4$ cm^{-2}) are achieved. Where the graded region extends further into the layer, the cross-hatch is not as clearly defined and the final EPD is much higher ($0.8–3\times10^5$ cm^{-2}). In the GMIRL device technology the monolith is taken from the centre of the uniform region where the EPD is low and the crystal composition is uniform.

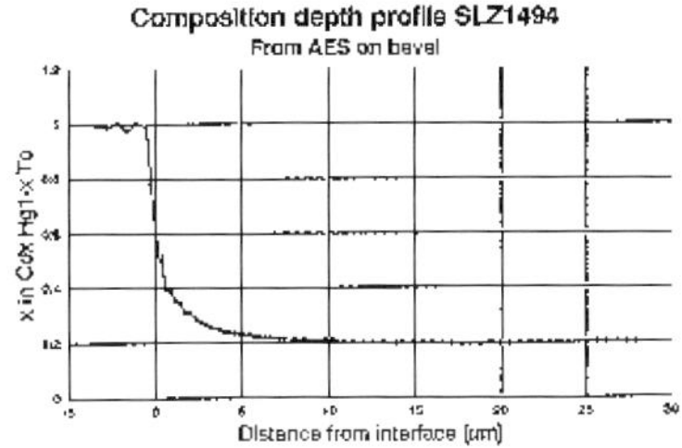

Figure 5. Composition through the thickness of an LPE layer by AES.

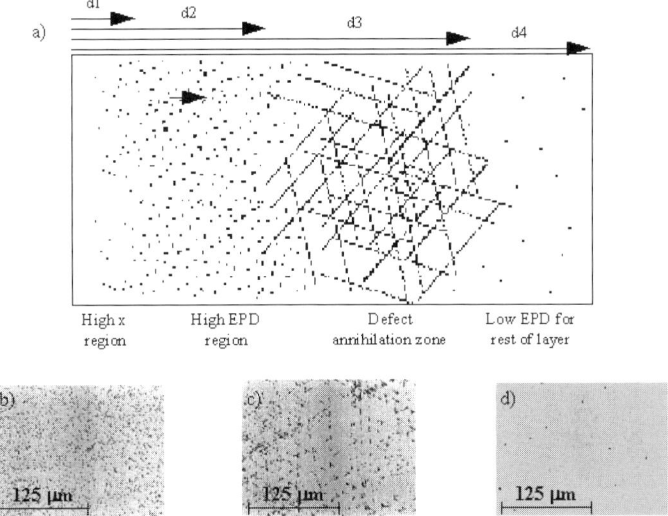

Figure 6. a) Schematic etch pit distribution on a low angle bevel, and actual micrographs b) close to the interface, c) in the graded region, d) in the uniform composition region.

INFLUENCE OF SUBSTRATE

When considering the choice of substrate material, lattice parameter match, crystal structure, survivability at the growth temperature, impurities, available size and cost are the principal concerns. Despite the relative difficulty of growing and processing these bulk crystal materials, members of the CdTe family are the material of choice. CdTe has been supplanted by CdZnTe as first choice due to its lower defect levels [4] (4×10^4 cm^{-2} compared to 1×10^5 cm^{-2}).

Substrates exhibit other properties which can lead to problems in growth. Precipitates of either Cd or Te greater than 10 μm in size cause pin-holes through the LPE layer, and twins, grain boundaries and micro-twins are replicated in the layer to give electrically active line features in FPAs. High IR absorption in the substrate makes IR spectroscopic assessment of the layer difficult, while surface and subsurface damage lead to poor nucleation and pin-holes. Non-flat growth surfaces (>5 μm) lead to excess retained melt after growth, and under or over-sized substrates can lead to sticking of the layer in the growth well. Variations in Zn concentration over substrate areas lead to differences in lattice matching and these variations are assessed prior to growth using near infrared spectrometry [5], see Figure 7.

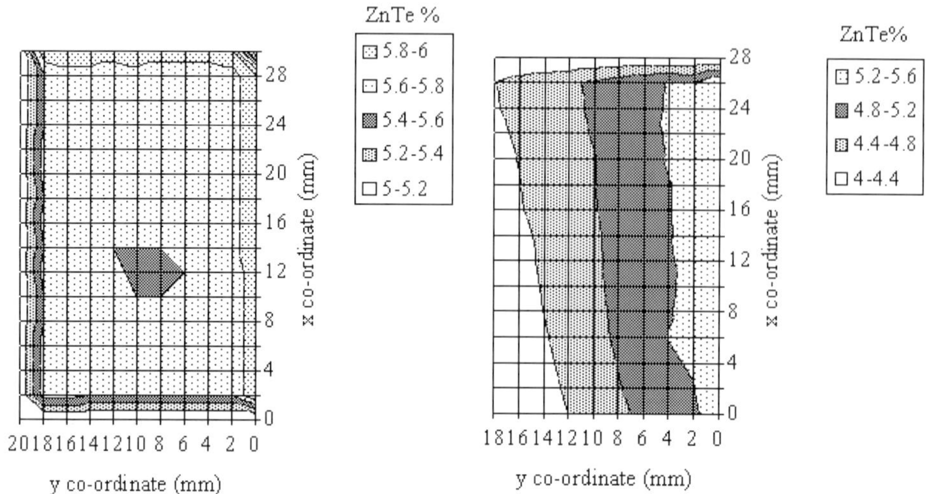

Figure 7. Zn distributions in 2 substrates showing small and large variations.

CONTROL OF METAL VACANCY LEVEL

Mercury vacancy levels in layers cooled directly from the growth temperature (500 °C) are too high for FPAs as they affect the junction forming behaviour. The required level is reached using Hg-rich annealing at ~300–400 °C for extended periods. Mapping the acceptor level, using 77 K Hall measurements, over entire layers shows uniformity of ±5% and sample to sample reproducibility of 90% within ±10% of the required level. This method is effective in controlling vacancy levels over the range 1×10^{16} to 3×10^{17} cm^{-3} [15]. Diode figures of merit, R_0A, (zero bias resistance ×diode area) are normally higher in LPE annealed using this process than in as-grown unannealed material.

BACKGROUND IMPURITY CONTROL

Mercury vacancies can be reduced by isothermal annealing of the CMT under a mercury atmosphere [7]. The electrical properties are then controlled by the presence of residual impurities. Considerable effort in determining which elements are electrically active in CMT [1, ch. 7, 5] has led to new chemical analysis techniques, such as laser scan mass spectrometry (LSMS) [8], but the sensitivity of these techniques is still not sufficient to measure impurities down to the 1×10^{14} cm^{-3} level. Nevertheless, Table I shows results of an LSMS analysis of a layer grown on an in-house produced substrate. This is a technique specifically developed to produce a survey of impurities in thin epitaxial layers. A Nd:YAG laser is rastered over the surface where it ablates material and produces a high yield of positive ions suitable for mass spectrometry. It has several important advantages over other mass spectrometry techniques, particularly with regard to CMT analyses, such as having relative ion sensitivities near unity, making calibration unnecessary. There is also a marked reduction in multiply charged ion species, thus reducing spectral interferences. Impurity levels, including O, in the centre of the layer are extremely low, as are those in the CdZnTe substrate, with the exceptions of C and O. There is also no evidence of interface concentration enhanced peaks which have been observed by other groups. This layer yielded a mid-10^{14} cm^{-3} n-type background level after Hg annealing at 200 °C.

Table I LSMS impurity survey in LPE layer grown on in-house CdZnTe substrate ($\times 10^{15}$ cm^{-3})

Element	Distance from surface (μm)						
	0–1	1–5	5–9	9–17	17–21	21–25*	25–27
C	1500	1000	3000	150	30	100	150
O	>300	15	5	1	1	1	10
Na	30	1.5	0.2	0.1	0.1	0.1	0.1
Si	600	1	1	1	<1	<1	<1
K	1.5	1	0.2	<0.03	<0.03	0.2	0.2

* Layer/substrate interface region.

High background donor levels (above about 10^{15} cm^{-3}) lead to poor electrical properties and donor levels above 4×10^{14} cm^{-3} are undesirable in photoconductive detectors. The background carrier levels in LPE CMT routinely exceed these demands, although potential donor elements, such as bromine are ever-present in laboratory environments and great care needs to be exercised to maintain low levels of this element.

SUMMARY

LPE from Te-rich solutions produces large areas of compositionally uniform CMT for use in large sized FPAs. The requirements of composition uniformity, thickness, defect density and electrical properties for FPAs have been met and the growth and annealing processes which produce the desired material described. Currently, growth is carried out on 2×3 cm and 3×4 cm CdZnTe substrates but scale-up to 4×6 cm is envisaged. Low background donor levels also make LPE layers desirable for use in photoconductive applications and good performance has been measured.

REFERENCES

1. P.Capper: (Ed) 'Narrow-Gap II-VI Compounds for Optoelectronic and Electromagnetic Applications', Chapman & Hall, London, 1997.
2. I.Gale, J.B.Clegg, P.Capper, C.D.Maxey, P.Mackett and E.S.O'Keefe: 'Measurement of CdHgTe composition depth profiles using Auger Electron Spectrometry on bevelled sections', *Adv. Mater. Opt. Electron.*, 1995, **5**, 79–86.
3. I.Hanert and M.Schenk: 'New etchants for CdTe and $Hg_xCd_{1-x}Te$', *J. Cryst. Growth*, 1990, **101**, 251–255.
4. P.Mackett: 'Defect etching of HgCdTe', in 'Properties of Narrow-Gap II-VI Compounds', ed. P.Capper, INSPEC/IEE, London, 1994, p.188–191.
5. P.Capper, E.S.O'Keefe, C.Maxey, D.Dutton, P.Mackett, C.Butler and I.Gale: 'Matrix and impurity distributions in CdHgTe(CMT) and (Cd,Zn)(Te,Se) compounds by chemical analysis', *J. Cryst. Growth*, 1996, **161**, 104–118.
6. D.Dutton, P.Capper, E.O'Keefe, P.Mackett, H.P.Hastings, S.Barton and C.L.Jones: 'Vacancy control of LPE and MOVPE grown CdHgTE by Annealing Techniques', *Mater. Sci. Forum*, 1995, **182-184**, 267–270.
7. D.Dutton, E.S.O'Keefe, P.Capper, C.L.Jones, S.Mugford and C.K.Ard: 'Type conversion of CdHgTe grown by Liquid Phase Epitaxy', *Semicond. Sci. Technol.* 1993, **8**, S266–S269.
8. F.Grainger, I.Gale, P. Capper, C.D.Maxey, P.Mackett, E.S.O'Keefe and J.Gosney: 'Impurity survey analysis of CdHgTe by laser scan mass spectrometry', *Adv. Mater. Opt. Electron.*, 1995, **5**, 71–78.

An Investigation into the Effect of Annealing on the Microstructure of CdTe/CdS Solar Cells

J.D.Painter, K.D.Rogers, and D.W.Lane
Department of Materials and Medical Sciences, Cranfield University, Shrivenham, Swindon, Wiltshire, SN6 8LA
Email j.d.painter@rmcs.cranfield.ac.uk, k.d.rogers@rmcs.cranfield.ac.uk, d.w.lane@rmcs.cranfield.ac.uk

Abstract

Thin film CdTe/CdS heterojunction solar cells are subjected to two anneal processes; the first after the deposition of the CdS, the second after the subsequent CdTe deposition. The effect of the second heat treatment on vacuum evaporated CdTe was investigated.

Samples were heat treated with and without a $CdCl_2$ surface coating at 450°C for various times. Recrystallization and interdiffusion occurred irrespective of the presence of $CdCl_2$, but $CdCl_2$ exacerbated the degree of interfacial mixing. The interdiffusion led to a reduction in the stress in the film, implying an accommodation of the interface mismatch, and thus a reduction in the number of defects. Similar interdiffusion characteristics were found when annealing at 400°C and 450°C, but interdiffusion was significantly reduced at the lower temperature of 350°C.

Introduction

CdTe is widely known as a promising low cost photovoltaic material, and is principally used in thin film devices due to its short optical absorption length. The most common CdTe solar cells are in the form of heterojunctions, with a CdS window layer.

CdTe has the ability to behave as an *n* or *p* type material, but is commonly deposited as *n*-type. As the CdS only behaves as *n*-type, the CdTe must be type converted by annealing the film in an atmosphere containing oxygen. The presence of $CdCl_2$ during this anneal is also necessary to produce a high efficiency solar cell. The $CdCl_2$ is believed to play an electrical role in the cell [1], and in the majority of cases it can also result in interdiffusion between the CdTe and CdS layers and the recrystallization of these layers, as indicated by a change in preferred orientation. These changes in the film structure are linked to the electronic properties of the cell. The interdiffusion between the layers is believed to reduce the strain at the interface, by accommodating the 10% lattice mismatch between the CdTe and CdS, thus reducing the number of defects. However, diffusion of Te into CdS reduces the high energy portion of the spectral response of the cell [2] lowering the current densities, and if the CdS is totally consumed the resulting $CdSTe/SnO_2$ contact is of inferior quality [3].

A previous study of depth profiling the cells revealed a two layered structure in the CdTe [4]. An as-deposited CdTe/CdS cell had small grains extending 0.6μm away from the interface, with larger grains on top. When annealed, the larger grains recrystallized forming a (220) texture, while the smaller grains underwent a partial recrystallization but remained with a (111) texture and incorporated sulphur diffusion from the CdS layer.

Experiment

CdS was deposited by the chemical bath method, as described elsewhere [5], onto commercially available tin oxide coated glass, with a film of thickness<0.1μm. Subsequently, the CdS films were annealed at 450°C in air for 15 minutes, and etched to remove any oxide from the film surface. The samples were immediately placed in the rotating sample holder of an Edwards Auto306 vacuum evaporator system for coating with CdTe. The chamber was evacuated to $9x10^{-6}$mB using a turbomolecular pump and liquid nitrogen cold trap. Prior to evaporation the substrates were heated to a temperature of 200°C. The source, initially covered by a shutter, was gradually heated until a stable deposition rate of $8nms^{-1}$ was acquired. The shutter was then withdrawn and the film thickness was monitored by a quartz crystal thickness monitor. The shutter was automatically closed when a predetermined thickness was reached. Prior to the second heat treatment, the films were dipped in a saturated $CdCl_2:CH_3OH$ solution and left to dry.

X-ray diffraction was performed on a Siemens D500 diffractometer employing CuKα radiation. The sample was continuously rotated during the data collection over an angular range of 22-50°. The XRD data was analysed by fitting the α_1 and α_2 components, but only the α_1 constituent was considered for further analysis. The degree of preferred orientation was calculated from the integrated intensity of the (220) and (111) Bragg maxima. The normalized ratio, R_n, was obtained by dividing the ratio of the heat treated sample by the ratio of a random sample, hence a randomly orientated film will have R_n=1.

$$R = A(220)/A(111) \quad (1)$$

$$R_n = R_{film}/R_{random} \quad (2)$$

Results

As-deposited CdTe has a pronounced (111) texture, Figure 1a. Determination of the lattice parameter revealed a 0.05% decrease, compared with the accepted value for pure CdTe. This decrease in a_0 along the growth axis indicates a slight strain which can be explained by a tensile stress perpendicular to the growth axis, i.e. in the growth plane. The decrease of just 0.05% is very small compared with the 10% lattice mismatch. However, as the diffraction pattern is an average of the electron density of the entire film, this can be explained by a thin strained layer at the interface, on top of which grows unstressed CdTe producing a diffracted average corresponding to a slightly strained film. Although the anneal may be expected to relieve the stress in the film, the difference in the thermal expansion coefficients between soda-lime glass and CdTe may actually introduce more stress into the film, creating defects and degrading the electrical performance of the cell [6].

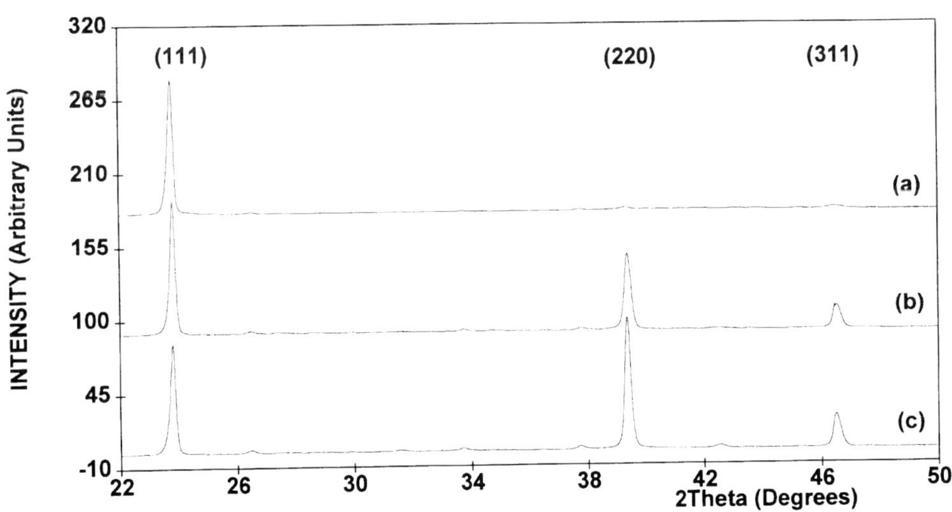

Figure 1 CdTe/CdS as-deposited (a) annealed with $CdCl_2$ at 450°C for 15 minutes (b) 30 minutes (c)

Two sets of samples were initialy heat treated at 450°C with and without $CdCl_2$, to investigate the role the $CdCl_2$ played. It was observed that the presence of $CdCl_2$ was not a requirement for recrystallization. Annealing samples with $CdCl_2$ results in the lattice parameter initially decreasing (Figure 2). This can be explained by the interdiffusion between the CdS and CdTe layers. The S is incorporated into the CdTe forming a $CdTe_{1-x}S_x$ layer by

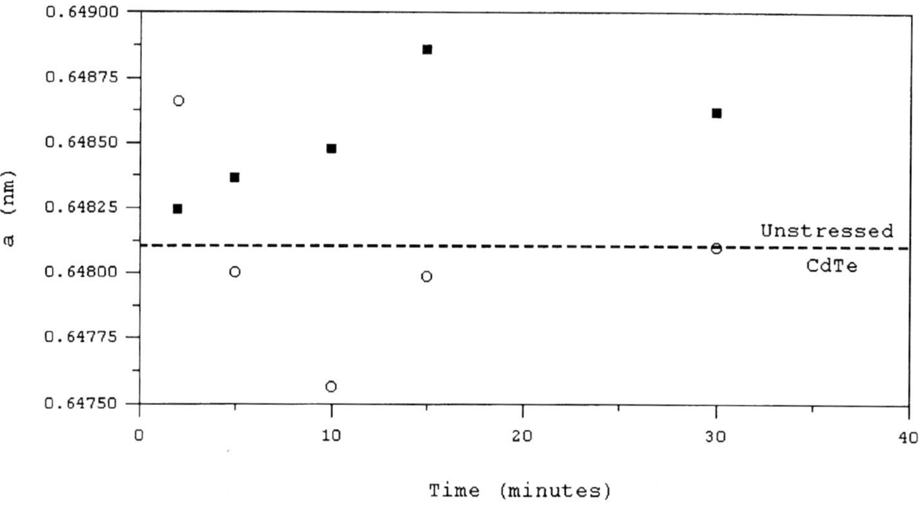

Figure 2 CdTe/CdS annealed at 450°C ○ with $CdCl_2$ ■ without $CdCl_2$

replacing Te atoms, and thus reducing the interatomic bond lengths and hence the size of the lattice, introducing compressive stress. After ~10 minutes the interdiffusion is complete and the lattice parameter then returns to a value approaching that of unstressed CdTe for longer anneal times. This indicates the dominant mechanism moves from interdiffusion to the formation, and perhaps growth, of stress free (S free) CdTe grains away from the interface. As stated earlier, a two layered structure in the CdTe film; large grain, stress free CdTe near the surface, small grain stressed $CdTe_{1-x}S_x$ near the interface with CdS, has been observed in previous work [4].

Without the $CdCl_2$ there was an initial sharpening of the Bragg maxima during the first five minutes of annealing, however the lattice spacing steadily increased to a value above that for pure CdTe, indicating increasing stress for intermediate times. The 30 minute anneal resulted in a decrease in the lattice parameter, indicating a reduction in the stress. This is inconsistent with the idea of stress being introduced into the CdTe by the different thermal expansion coefficient of the glass. Therefore, another factor must be involved, possibly limited interdiffusion across the interface. McCandless et al [2] have previously reported a small amount of interfacial mixing with no $CdCl_2$ present.

Figure 3 plots the normalized integrated intensity ratio of the (220) peak against that of the (111) peak to gauge the degree of recrystallization. The sample annealed following the $CdCl_2$ treatment initially recrystallizes to an almost random orientation, then (220) texture develops (this is similar behaviour to that found in electrodeposited CdTe). This is consistent with the model of grain growth in the upper (220) textured CdTe layer with longer anneal times. The sample annealed without the $CdCl_2$ treatment initially recrystallizes to a greater (220) texture than the $CdCl_2$ treated sample, but after thirty minutes the positions are reversed.

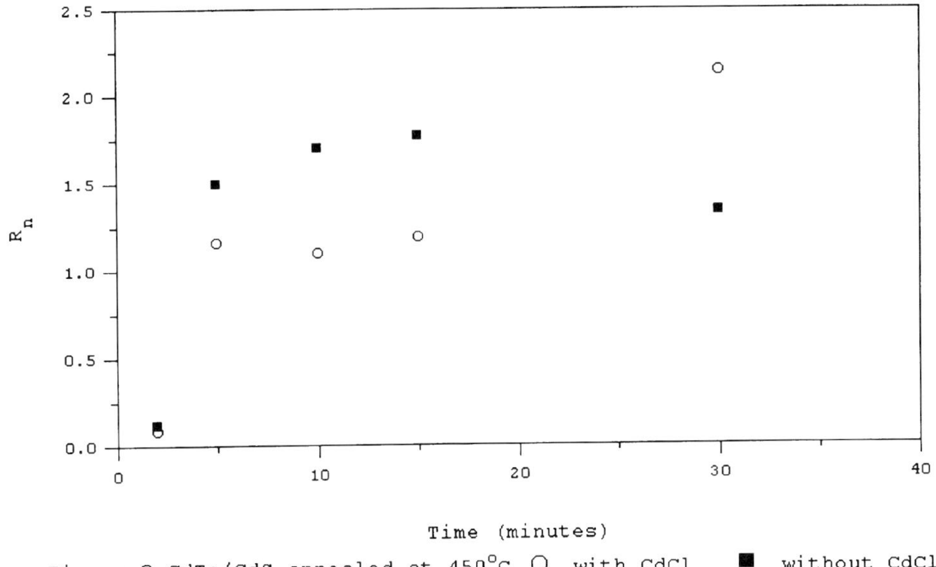

Figure 3 CdTe/CdS annealed at 450°C ○ with $CdCl_2$ ■ without $CdCl_2$

A further study of CdTe recrystallization with $CdCl_2$ was conducted at different temperatures, upon which a minimum temperature of 350°C was found to be necessary. Figure 4 shows R_n at different temperatures and times. The change in the lattice parameter (Figure 5) reveals the similar behaviour (interdiffusion and recrystallization/grain growth) of samples annealed at 400°C and 450°C, albeit at a slower rate. Fitting the behaviour of the sample annealed at 400°C to the model previously described, would indicate that the CdTe film did not undergo the same scale of grain growth. At 350°C the recrystallization results in an almost random orientation, but the interdiffusion is more limited and the stress in the film is not fully removed.

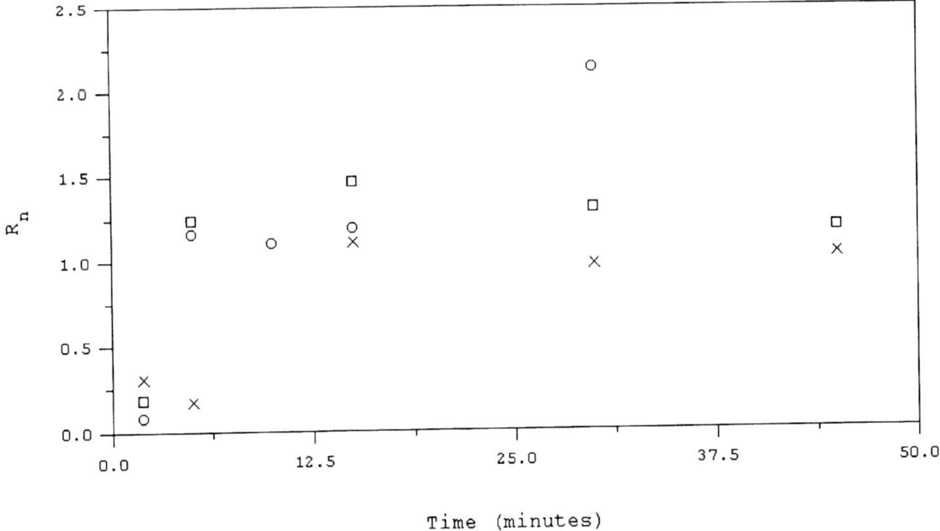

Figure 4 CdTe/CdS annealed with $CdCl_2$ at ○ 450°C □ 400°C × 350°C

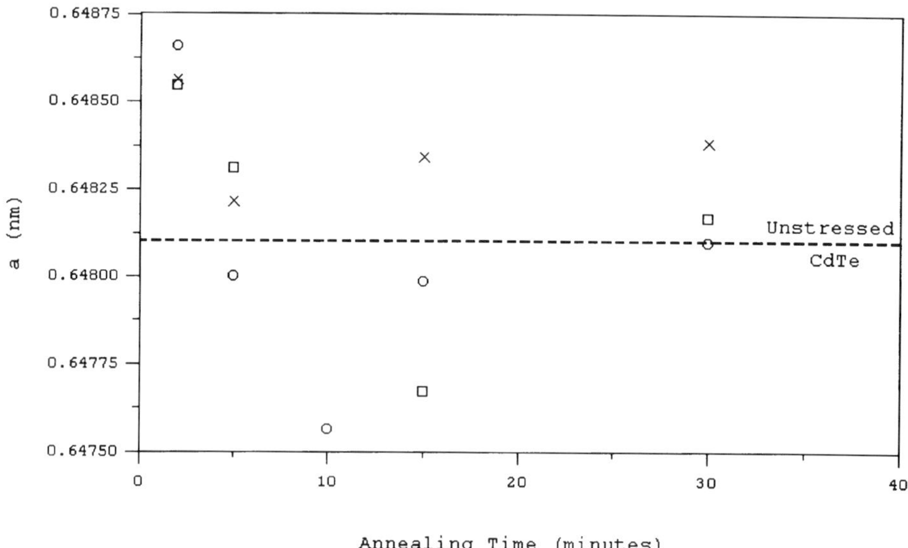

Figure 5 CdTe/CdS annealed with $CdCl_2$ at ○ 450°C □ 400°C × 350°C

Conclusion

The as-deposited CdTe film consists of single phase, slightly strained material due to the lattice mismatch. Heat treatment of the sample without the $CdCl_2$ treatment initially results in a strained, (220) textured film for intermediate anneal times. The stress in the film must be relieved during recrystallization, but the different thermal expansion coefficients reintroduce stress upon cooling. Longer anneals reduce both the texture and the strain due to a small amount of interfacial mixing. The recrystallization of CdTe without $CdCl_2$ is unusual, it does not occur with most deposition techniques, but it has previously been observed in other vacuum evaporated CdTe films [2]. This suggests major differences in the CdTe film structure and/or composition which warrants further investigation.

Annealing with $CdCl_2$ at 450°C, results in a two stage process. Interdiffusion occurs over ~10 minutes forming a $CdTe_{1-x}S_x$ layer, with the film having a random orientation. Once the interdiffusion is complete the emphasis shifts to the production of nearly stress free CdTe grains, probably including grain growth. At 400°C, a similar process occurs over a slightly longer time frame. Annealing at 350°C results in recrystallization but the interdiffusion occurs over a much smaller scale, hence the stress is not completely relieved.

Acknowledgements

This work was sponsored by EPSRC CASE award contract number GR/9539270X. Thanks to Dr M E Ozsan at BP Solar for financial support and for providing the CdS samples.

References

1. A.NIEMEGEERS, M.BURGELMAN, H.RITCHER and D.BONNET: 'A Simple Model for the Effects of the $CdCl_2$ Treatment on the Performance of CdTe/CdS Solar Cells', Proc. 14th European PVSC, 1997, 2079-2082

2. B.E.McCANDLESS, L.V.MOULTON and R.W.BIRKMIRE: 'Recrystallization and Sulfur Diffusion in $CdCl_2$-Treated CdTe/CdS Thin Films', Prog. Photovolt. Res. Appl., 1997, **5**, 249-260

3. B.E.McCANDLESS and S.S.HEGEDUS: 'Influence of CdS Window Layers on Thin Film CdS/CdTe Solar Cell Performance', Proc. 22nd IEEE PVSC, 1991, 967-972
4. K.D.ROGERS, J.D.PAINTER, M.J.HEALY, D.W.LANE and M.E.OZSAN: 'The Crystal Structure of CdTe-CdS Thin Film Heterojunction Solar Cells', *To be published in Thin Solid Films*
5. R.ORTEGA-BORGES and D.LINCOT: 'Mechanism of Chemical Bath Deposition of Cadmium Sulfide Thin Films in the Ammonia-Thiourea System', J. Electrochem. Soc., 1993, **140** (12), 3464-3473
6. S.JIMENEZ-SANDOVAL, M.MELENDEZ-LIRA and I.HERNANDEZ-CALDERON: 'Crystal-Structure and Energy-Gap of CdTe Thin-Films Grown by Radio-Frequency Sputtering', J. Appl. Phys., 1992, **72** (9), 4197-4202

CARRIER DENSITY MICROSCOPY OF CdS/CdTe SOLAR CELLS

P. R. EDWARDS*, S. A. GALLOWAY**, K. DUROSE*
* Department of Physics, University of Durham, South Road, Durham DH1 3LE, UK
E-mail p.r.edwards@durham.ac.uk, ken.durose@durham.ac.uk
** Oxford Instruments, Tubney Woods, Abingdon, Oxon OX13 5QX, UK
E-mail simon.galloway@oxinst.co.uk

ABSTRACT

A novel technique is proposed for probing variations in carrier density in semiconductor devices on a microscopic scale. The method uses the electron beam-induced current (EBIC) mode of a scanning electron microscope, which evaluates spatial variations in the carrier collection efficiency of a semiconductor junction. Increasing the electron beam current, and hence the injected excess carrier density, has been seen to result in a decrease in EBIC efficiency. This effect is described in terms of the onset of high injection conditions, and measuring the current threshold at which this occurs allows the estimation of equilibrium carrier densities. The technique has been applied to polycrystalline CdS/CdTe solar cells, and the results show grain-to-grain variation as well as higher carrier densities at grain boundaries. Comparison with results from control samples lacking a CdS layer indicate that sulphur diffusion increases the density of electrically active levels in the CdTe.

INTRODUCTION

The cadmium sulphide/cadmium telluride solar cell is one of a number of emerging technologies in the field of low-cost thin-film photovoltaics. It consists of layers of polycrystalline CdTe and CdS deposited on a substrate of transparent conducting oxide (TCO) coated glass (see figure 1). Light penetrates the thin window layer of wide bandgap CdS to be absorbed the thicker CdTe layer, and the resultant charge carriers are collected by the *p-n* junction.

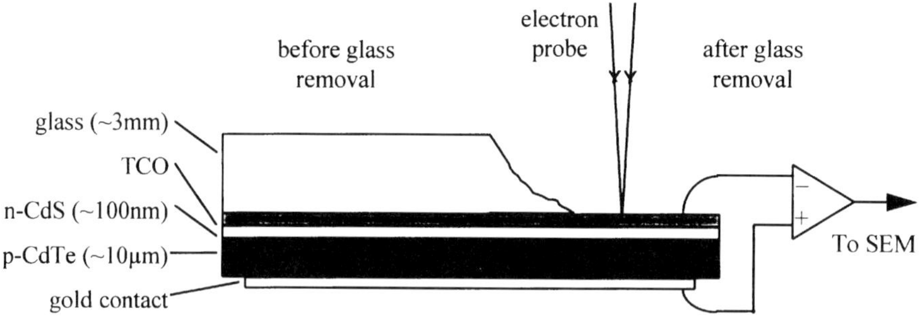

Figure 1: Structure of CdS/CdTe solar cell and EBIC set-up

Although small area cells have been produced with efficiencies up to 16%[1,2], this has only been made possible by the use of a well known but poorly understood post-deposition treatment. The details of this treatment vary for different manufacturers and deposition techniques, but all involve the introduction of chloride ions followed by annealing in an oxygen-containing atmosphere. This process can result in an increase in cell efficiency of up to an order of magnitude, and is thought to be at least partly due to passivation of the grain boundaries.

A further issue of interest is that of the diffusion of sulphur from the CdS layer to the CdTe layer, either during deposition or during post-deposition annealing. This forms an interfacial layer of the alloy CdS_xTe_{1-x} with a lower bandgap than CdTe, which improves the spectral response of the cell. Evidence for this diffusion has been found by several techniques[3,4], and it has been reported that chloride treatment enhances the effect[3].

THEORY

Electron beam-induced current (EBIC) measurements are widely used in semiconductor physics to measure diffusion lengths and to probe minority carrier recombination. The technique involves first creating a depletion region in the sample by making either a p-n junction or Schottky barrier. An electron beam is then used to generate electron-hole pairs in the semiconductor by the process of impact ionisation. The carriers are collected by the electric field, and the resultant current I_{EBIC} measured. This is often normalised to take into account the electron beam power, giving an EBIC efficiency[5], η:

$$\eta = \frac{1}{(1-f)} \cdot \frac{I_{EBIC}\varepsilon_{e-h}}{I_b E_b} \qquad (1)$$

where I_b and E_b are the electron beam current and energy respectively, f is the fraction of primary electrons that are backscattered, and ε_{e-h} is the energy required to generate one electron-hole pair (approximately $3\times E_g$). In practice, the measured value of I_{EBIC} is subject to losses incurred by the series resistance of the sample. This means that the EBIC signals can only be compared relative to others from the same sample, and not to absolute values.

When the beam current is low, the density of injected excess carriers is not significant compared to the number of equilibrium carriers, and the device is said to be in the low injection regime. This corresponds to the normal working conditions of the solar cell under unconcentrated sunlight. When the number of injected carriers becomes comparable to or exceeds the equilibrium value, medium or high injection conditions are reached. Under these conditions, the presence of the extra carriers acts to lower the value of the electric field in the probed area. This reduces the probability of collection for a given carrier, and hence reduces the EBIC efficiency. If the carrier injection density at which this phenomenon occurs can be determined experimentally, then information can be inferred about the equilibrium carrier density.

EXPERIMENT

The samples used in this study were fabricated by ANTEC GmbH using the close-space sublimation method[6]; both semiconductors were deposited with a substrate temperature of 500°C. In order to study the effect of the post-deposition treatment, three differently treated samples were prepared:-

a) as-deposited
b) 60nm layer of $CdCl_2$ evaporated on to CdTe, followed by annealing in air at 400°C for 30 minutes
c) control sample without CdS layer, in order to characterise the effect of sulphur diffusion across the interface

The glass substrate was then removed by mechanical polishing followed by etching with hydrofluoric acid, the TCO acting as a stop-layer. This was to allow the electron beam to enter the solar cell through the front-wall, in order to probe the junction region, the area of most relevance to device operation. Quantitative EBIC images were recorded in a JEOL JSM-IC848 scanning electron microscope using a Matelect ISM5 amplifier. The electron beam energy was kept constant at 11 keV, to provide a depth-dependant carrier generation function as close as possible to that of white light. Injection-dependent effects were then investigated by varying the beam current over five orders of magnitude.

RESULTS AND DISCUSSION

Some of the results of EBIC measurements on the untreated solar cell are seen in figure 2. (As with all the EBIC images shown here, the brightness and contrast have been adjusted to allow the details to be seen.)

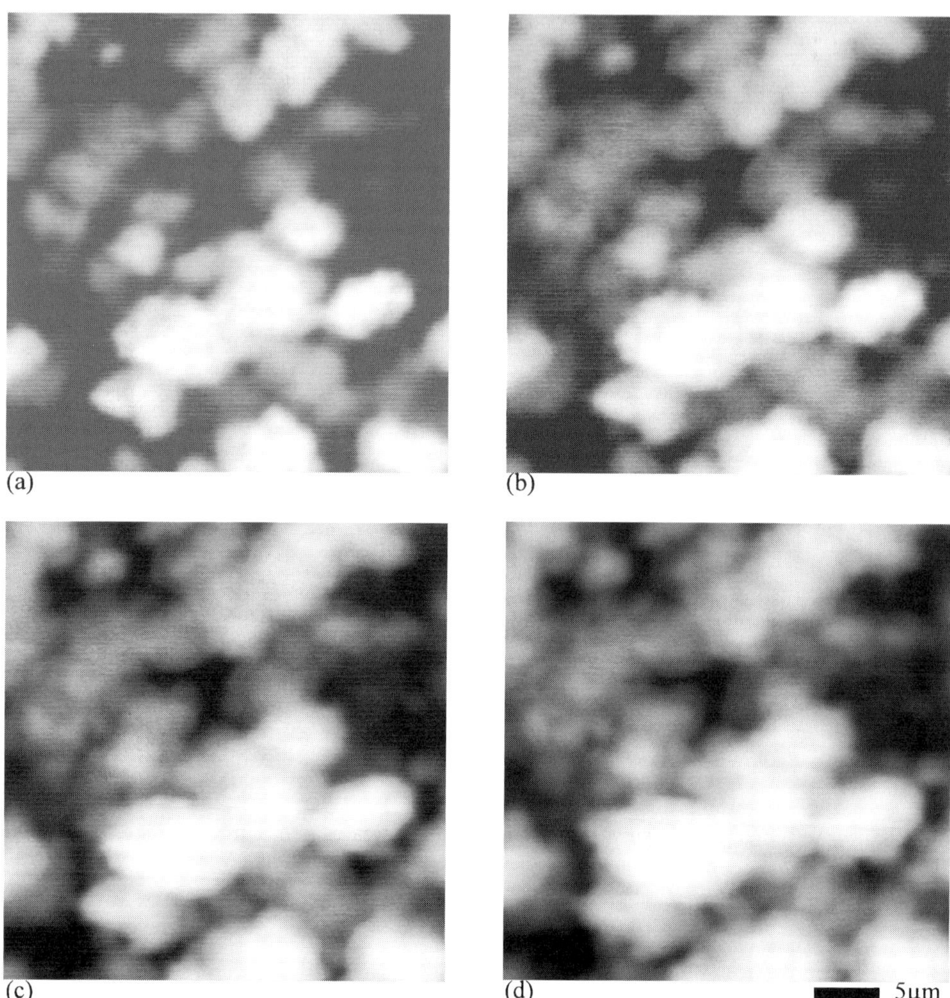

Figure 2: EBIC images of a similar area of an untreated cell, taken with varying primary beam currents:-
(a): 1×10^{-10} A (b): 5×10^{-10} A (c): 2×10^{-9} A (d): 1×10^{-8} A

It can be seen that although some grains are bright (corresponding to high current collection), there are also large areas giving little or no EBIC. This may be due to variation in the depth of the collecting junction, which would result from inter-grain variation in doping density.

EBIC images of the chloride-treated device (some of which are shown in figure 3) show a greater variation with beam current:

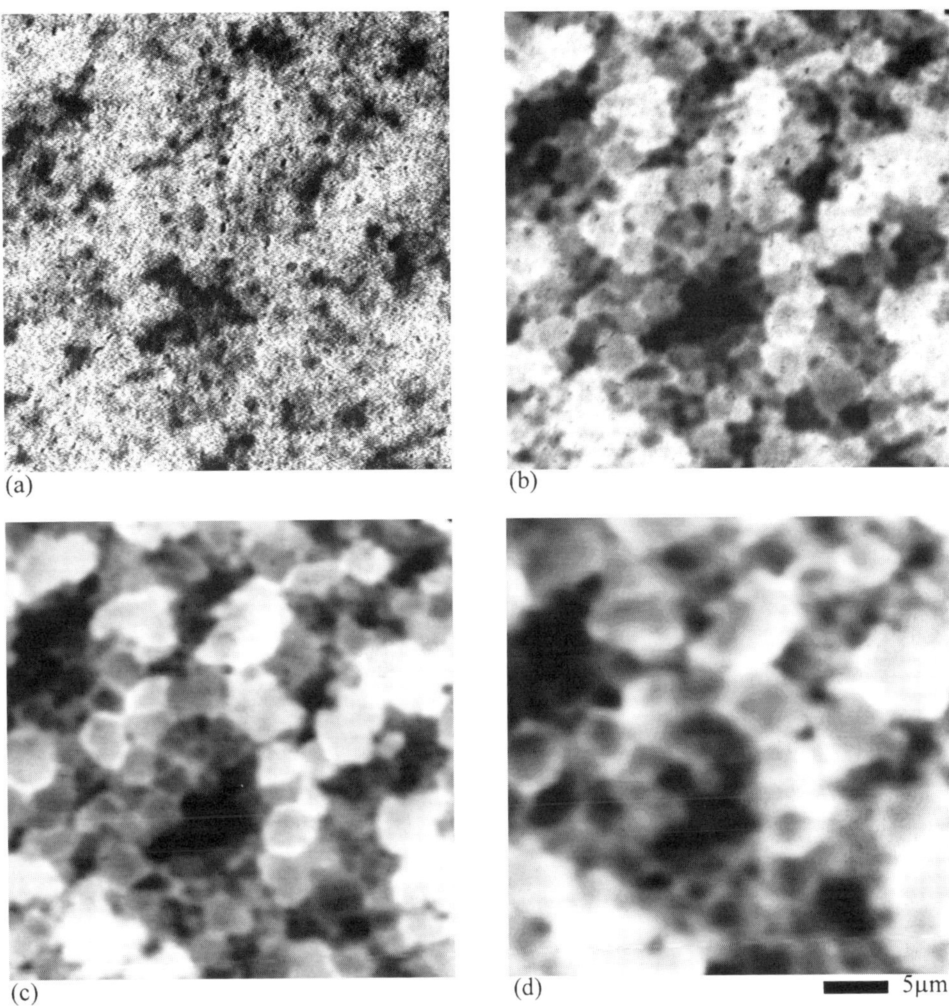

Figure 3: EBIC images of a similar area of a chloride-treated cell with CdS, taken with varying primary beam currents:-
(a): 1×10^{-10} A (b): 5×10^{-10} A (c): 2×10^{-9} A (d): 1×10^{-8} A

a) At the relatively low beam current of 10^{-10}A, there is no grain boundary contrast, although some difference is seen between grains. At this stage, all areas are under low injection.
b) Increasing to 5×10^{-10}A, some of the grains begin to show bright grain boundaries, as the grain interiors start to move to high injection conditions but the (more highly doped) boundary regions remain in low injection.
c) As the beam current is increased still further, this effect becomes more pronounced, with more grains becoming darker and so enhancing the bright contrast at the boundaries.
d) By 10^{-8}A, some of the grain boundaries are beginning to become darker, as the density of injected carriers becomes high enough for the boundary regions to reach high injection.

Figure 4 shows some of the EBIC results for the CdS-free control sample:

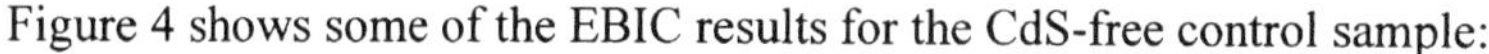

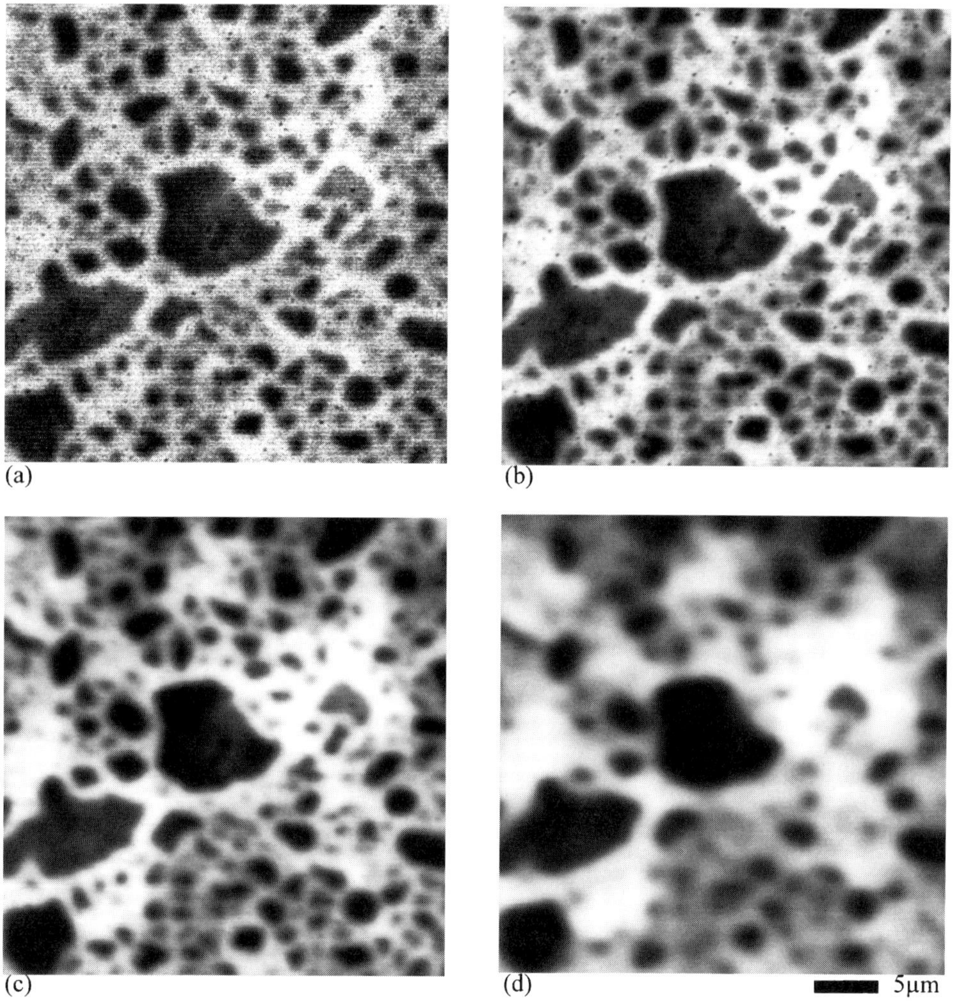

(a) (b)

(c) (d) 5μm

Figure 4: EBIC images of a similar area of a chloride-treated cell without CdS, taken with varying primary beam currents:-
(a): 1×10^{-10} A (b): 5×10^{-10} A (c): 2×10^{-9} A (d): 1×10^{-8} A

a) At a beam current of only 10^{-10}A, the sample is already showing bright boundary contrast (compare this with figure 3a for the sample with CdS). From this it is inferred that the grain interiors may have lower impurity concentrations than those in the corresponding cell with CdS, so that high injection conditions have already been reached within the grain.

b) to d) Some decrease in the bright boundary contrast may be seen as the beam current is increased, but this is complicated by the loss of spatial resolution inherent when more carriers are injected.

A quantitative analysis of the data in figures 3 and 4 has been carried out by considering the change in EBIC efficiency with beam current *pixel by pixel*. (It was not possible to use the same method on the untreated cell (figure 2), as there was insufficient change in EBIC with increasing injection.) As an example, figure 5 shows the EBIC efficiency versus beam current for two different grains from the images in figure 4.

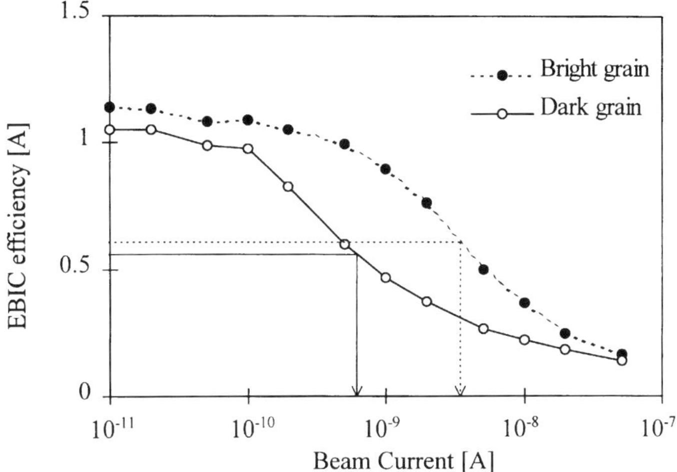

Figure 5: EBIC efficiency as a function of electron beam current for two different features on the sample

It can be seen from these graphs that the EBIC efficiency is relatively insensitive to changes in beam current at both high and low beam injection extremes, but changes significantly around the medium injection regime. This occurs under different injection levels for the two grains, and it is this effect that was exploited to estimate the equilibrium carrier levels. Also shown on the graph is the method that was used to define the threshold current, indicated by the arrowed lines. A beam current was interpolated from the graph at a point half-way between the low-injection and high-injection EBIC efficiency values. (This figure of half was chosen arbitrarily.) This process was carried out *for each pixel*, and the resultant threshold values were then reassembled to produce a current map, as shown in figure 6.

In these images, bright areas correspond to regions with a high beam current thresholds, and hence (it is inferred) high equilibrium carrier densities. However, the large number of complicating factors involved prevent the unambiguous interpretation of these results. The depth-dependence of the collecting junction (thought to be a buried homojunction within the CdTe rather than the actual CdS/CdTe interface) could have an effect on the EBIC images, although this would not be strongly injection-dependent. This would explain the large areas of low collection seen in the untreated sample (figure 2). A further influence would be that of the spatial dependence of the carrier lifetimes; this is of particular importance at the grain boundaries, where preferential recombination has previous been reported as the cause of dark grain boundary contrast seen using back-wall EBIC on similar devices[7].

The distribution histograms of calculated threshold currents for the two samples can be seen in figure 7. From these it is observed that one effect of the CdS layer is to increase the overall values of the threshold. This can explained by the diffusion of sulphur into the CdTe, increasing the concentration of impurities.

No attempt is made here to quantify the carrier injection densities associated with the measured threshold currents. This would require not only an estimation of the carrier generation function of the electron beam (either by use of an analytical function or a Monte Carlo simulation), but also the solution to the drift and diffusion equations for the injected electrons and holes.

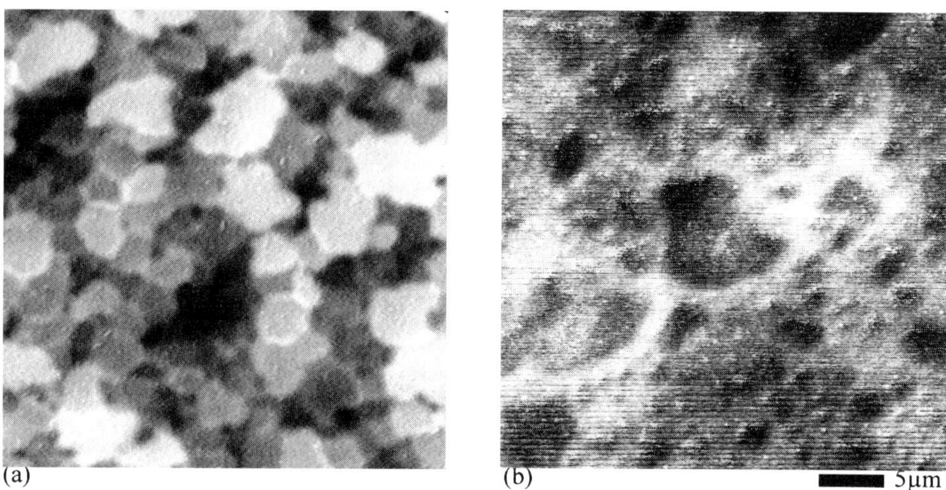

Figure 6: Map of current threshold values for (a) chloride treated cell *with* CdS layer and (b) chloride treated cell *without* CdS layer. These maps were constructed from the EBIC images seen in figures 3 and 4 and further EBIC images taken with different beam currents (not shown). Black = 1nA, white = 4 nA beam current.

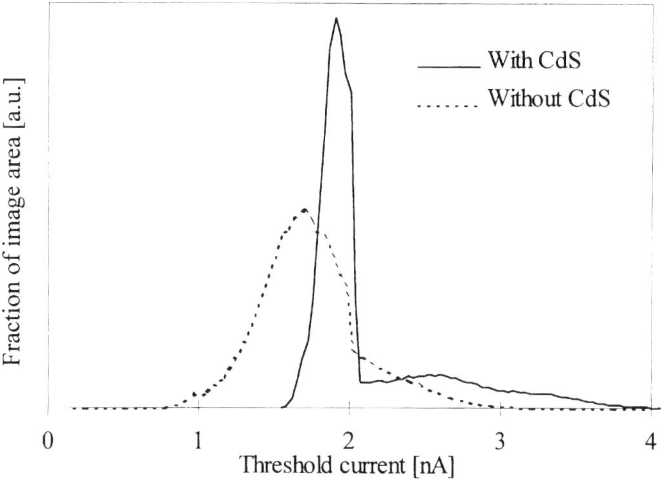

Figure 7: Histograms showing distribution of threshold beam currents for cells with and without cadmium sulphide layers.

CONCLUSIONS

It has been demonstrated that useful information can be obtained by considering the beam current dependence of EBIC images. This has been carried out on CdS/CdTe solar cells, with the result that it is now clear that the grain boundary contrast usually seen in EBIC images of these structures cannot wholly be attributed to grain boundary recombination. Rather, grain boundary regions have been shown to have higher equilibrium carrier densities than the grain interiors, suggesting that impurity transport during post-deposition processing is dominated by the boundaries. The presence of cadmium sulphide in the solar cell has also been seen to increase carrier density in the device. Determination of absolute carrier concentrations, however, has not been possible, due to the contribution of other factors; these include the possibility of spatially dependent carrier lifetimes, and the dependence of the EBIC signal on the depth of the collecting junction. Unambiguous interpretation of these results is further complicated by the need to compare EBIC images of different resolution, as this resolution is also dependent upon injection conditions.

ACKNOWLEDGEMENTS

The authors would like to thank D. Bonnet and H. Richter at ANTEC GmbH for providing the solar cells and for valuable discussions. Paul Edwards gratefully acknowledges the financial support of EPSRC.

REFERENCES

1. J.BRITT and C.FEREKIDES: 'Thin-film CdS/CdTe solar cell with 15.8% efficiency', *Appl. Phys. Lett.*, 1993, **62** (22), 2851-2852.
2. T.ARAMOTO, S.KUMUZAWA et al.: '16% efficient thin-film CdS/CdTe solar cells', *Jpn. J. Appl. Phys.*, 1997, **36**, 6304-6305.
3. B.E.McCANDLESS, L.V.MOULTON and R.W.BIRKMIRE: 'Recrystallization and sulfur diffusion in $CdCl_2$-treated CdTe/CdS thin films', Progress in Photovoltaics, 1997, **5** (4), 249-260.
4. P.R.EDWARDS, D.P.HALLIDAY and K.DUROSE: 'The influence of $CdCl_2$ treatment and interdiffusion on grain boundary passivation in CdTe/CdS solar cells', *Proc. 14th European Photovoltaic Solar Energy Conference*, 1997, **II**.
5. N.TABET and R.-J.TARENTO: 'Calculation of the electron-beam-induced current (EBIC) at a Schottky contact and comparison with Au/n-Ge diodes', *Phil. Mag. B*, 1989, **59** (2), 243-261.
6. D.BONNET, H.RICHTER and K.-H.JÄGER: 'The CTS thin film solar module - closer to production', *Proc. 13th European Photovoltaic Solar Energy Conf.*, 1995, **II**, 1456-1461.
7. S.A.GALLOWAY, A.J.HOLLAND, et al.,'Microscopic characterisation of CdS/CdTe thin film photovoltaic devices using scanning optical and electron beam injection techniques', *Proc. 13th European Photovoltaic Solar Energy Conf.*, 1995, **II**, 2072-2075.

PREPARATION AND HUMIDITY SENSITIVE IMPEDANCE OF THE SPINEL CERAMIC Ni_2GeO_4

M. J. Hogan*, A. W. Brinkman, T. Hashemi, M. A. Cousins
Dept. of Physics, University of Durham, DH1 3LE, U.K.
*email: m.j.hogan@durham.ac.uk

ABSTRACT

Humidity dependant resisitivity has been reported in several ternary oxide ceramics with a spinel crystal structure, a number of which have been investigated for potential use as humidity sensors. The spinel ceramic material nickel germanate, Ni_2GeO_4, has been found by us to possess a humidity dependant impedance, exhibiting a resistance change of several orders of magnitude over the humidity range 10% to 90% at 25 °C. This paper reports on the preparation and humidity sensitive properties of the material.

Nickel germanate has been prepared by solid state reaction from nickel (II) oxide and germanium (IV) oxide precursors, intimately mixed in the molar ratio 2:1. Samples of the unreacted precursors have been calcined at various temperatures up to 1500 °C in air, and have subsequently been investigated by powder x-ray diffraction (XRD) and infra-red spectroscopy (FTIR). These techniques show that the reaction starts at around 900 °C and that single phase material can be produced by firing the precursor mixture at 1200 °C for 12 hours.

Pellet samples of nickel germanate have been prepared from single phase material sintered at 1200 °C. The humidity dependant impedance has been investigated at 25 °C over a range of humidity between 22% and 100% R_H, by impedance spectroscopy in the frequency range 10^{-2} to 10^7 Hz. The d.c. resisitivity of the samples typically varies from $\sim 10^{11}$ Ωcm to $\sim 10^5$ Ωcm as the ambient humidity is changed from 10 %R_H to 90 %R_H at 293 K.

INTRODUCTION

In recent years, there have been many investigations into the use of porous ceramic materials as humidity sensors [1, 2], since they offer significant advantages over polymer thin film sensors in terms of resistance to chemical attack, and their thermal and physical stability [3]. Nickel germanate, Ni_2GeO_4, is such a polycrystalline ceramic material, with a spinel-type crystal structure, which has been found by us to demonstrate humidity dependant impedance, and thus is therefore a potential candidate for use as a humidity sensor. Previous investigations of Ni_2GeO_4 have been confined to its formation [4-7], optical properties [8, 9], and crystal structure [10].

EXPERIMENTAL

High purity (>99.99%) powders of nickel (II) oxide and germanium (IV) oxide, in the molar ratio 2:1, were intimately mixed in a pestle and mortar to obtain a precursor material. This mixture was placed in a high purity alumina boat (Multilab Alsint 99,7), and calcined at varying temperatures between 700 °C and 1400 °C in a tube furnace (Lenton Thermal Designs) open to the atmosphere. X-ray diffraction (Philips PW1130 diffractometer, λ = 1.5406 Å) and infra-red spectroscopy (Bruker IFS48, KBr disc samples) were used to characterise the samples at each stage in the fabrication.

Monophasic Ni_2GeO_4 powder formed at 1200 °C was then reground, pressed at 2 tons cm^{-2} into 13 mm diameter pellets of approximately 1 mm thickness, and subsequently fired at 1200 °C. Gold contacts were sputtered onto the faces of these pellets, using an argon plasma, in order to provide electrical connection to the sample.

The d.c. resistance of the samples was measured by a two-point probe technique using a current source and digital electrometer (Keithley 617), with the sample humidity and temperature controlled by a computer-driven chamber, over the range 20 $\%R_H$ to 98 $\%R_H$ at a constant temperature of 293 K, as measured by a Rotronic AM3-HP100A polymer thick film sensor.

The ambient humidity and temperature of the samples for a.c. measurements was controlled in a Hereaus environmental chamber, utilising dew-point mirror control of humidity over the range 20 $\%R_H$ to 98 $\%R_H$ at a constant temperature of 298 K. The a.c. impedance measurements were carried out using a Solartron 1296 Impedance Analyser, interfaced to the sample through a Solartron 1295 Dielectric Interface, over the frequency range 10 mHz to 10 MHz.

RESULTS AND DISCUSSION

Calcination

X-ray powder diffraction spectra taken from samples of the ($NiO + GeO_2$) precursor fired at different temperatures only displayed peaks associated with Ni_2GeO_4 and unreacted precursor. No other phases were detected.

Quantitative phase analysis has been performed on these x-ray results as shown in figure 1, where the concentration shown is calculated according to

$$I_1 / I_S = \text{const.}\ x_1 \qquad (1)$$

where the ratio I_l / I_S is the ratio of the intensity of component 1 to the internal standard, which in this case was 50 wt.% Si, and x_l is the molar fraction of component 1 in the mixture [11].

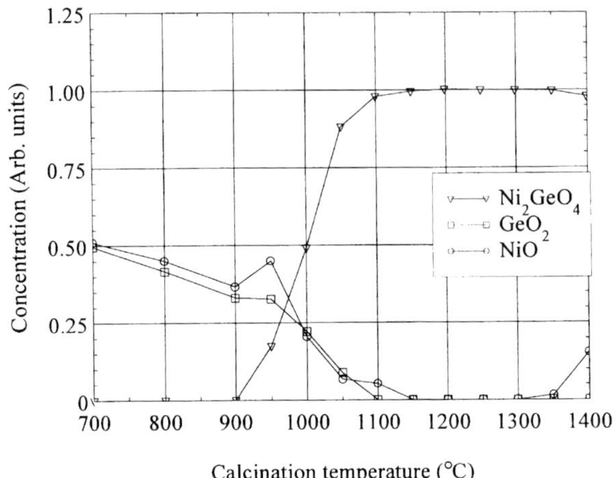

Figure 1 Quantitative phase analysis of XRD of the reaction ($2NiO + GeO_2 \rightarrow Ni_2GeO_4$)

This graph shows the progression of the reaction {$2NiO + GeO_2 \rightarrow Ni_2GeO_4$} with temperature, for a firing time of 12 hours. The conversion of the precursor materials into the product is clearly visible. At temperatures less than 900°C, there was no detectable production of Ni_2GeO_4, and the concentrations of the precursors steadily decreased with temperature as they were converted into Ni_2GeO_4. From figure 1, we see that detectable production of Ni_2GeO_4 started at 950 °C, the precursors were exhausted by 1150 °C indicating that monophase material had been formed since no other phases were observed, and that decomposition of Ni_2GeO_4 involving production of NiO started to occur at temperatures above ~1350 °C.

These results were confirmed by FTIR analysis of the powders, as shown in figure 2. FTIR spectra of four samples are shown (displaced along the vertical axis for clarity), the precursor powder, and specimens of the precursor powder which had been fired at 700 °C, 900 °C, and 1200 °C. In the trace corresponding to the unfired precursor powder, there are two main features, a strong absorption at $\nu = \sim 880\ cm^{-1}$ due to a GeO_2 bond vibration, and a wide absorption band between $\nu = \sim 600\ cm^{-1}$ and $\nu = \sim 400\ cm^{-1}$, of which GeO_2 is responsible for absorption between $\nu = \sim 600\ cm^{-1}$ and $\nu = \sim 500\ cm^{-1}$, and NiO is responsible for absorption between $\nu = \sim 500\ cm^{-1}$ and $\nu = \sim 400\ cm^{-1}$. The top trace, which is due to pure Ni_2GeO_4, shows two strong absorptions, one at $\nu = \sim 670\ cm^{-1}$, and another at $\nu = \sim 440\ cm^{-1}$. However, for the purposes of this analysis, only the absorption at $\nu = \sim 670\ cm^{-1}$ shall be considered, since it does not correspond to any absorptions of the precursor mixture.

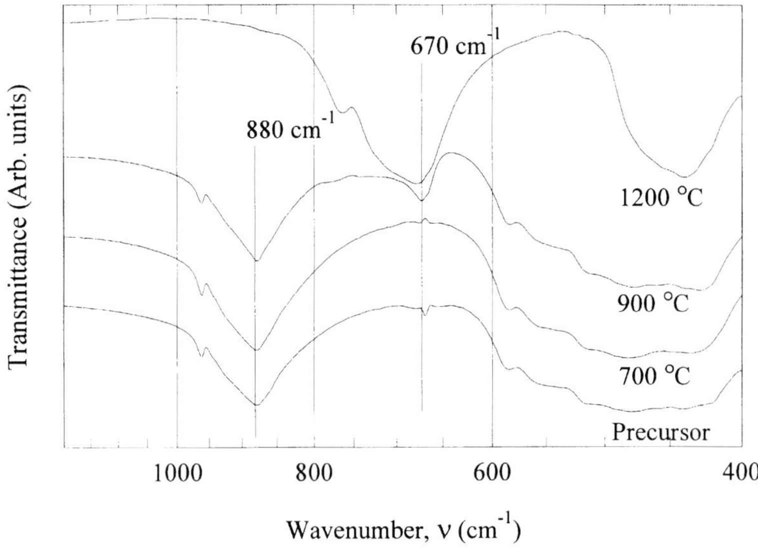

Figure 2 FTIR Spectra of precursor powder calcined at 1200 °C, 900 °C, 700 °C and as prepared

The FTIR trace produced by the sample calcined at 700 °C was essentially identical to that produced by the precursor material, showing that no conversion into the desired product had taken place. However, at 900 °C, there was a clearly visible absorption at $\nu = \sim 670$ cm^{-1}, which was indicative of Ni_2GeO_4 formation. There were no signs of the absorptions due to GeO_2 or NiO, in the FTIR spectra of material fired at 1200°C, indicating this material was single phase.

Nickel germanate formation was detected at a lower temperature using the FTIR technique than the XRD technique, 900°C for FTIR as compared to 950 °C for XRD, due to the higher sensitivity of FTIR to small concentrations of material.

Sintering

Upon firing at 1200 °C, the pressed pellets underwent a lateral shrinkage of (2.1±0.5)%, indicating that sintering of the ceramic grains had taken place. Density measurements of the sintered pellets revealed an average density of 3.2×10^3 kg m^{-3}, compared to a theoretical single crystal density of 6.1×10^3 kg m^{-3} [12], implying a total porosity (open and closed) of 48%. Image analysis of scanning electron micrographs gave an average grain size of (0.51±0.2)μm. X-ray powder diffraction performed on the sintered pellets confirmed that the material was still monophasic.

Electrical Properties

The d.c. resistivity of the pellet samples was observed to change approximately exponentially with humidity as per figure 3.

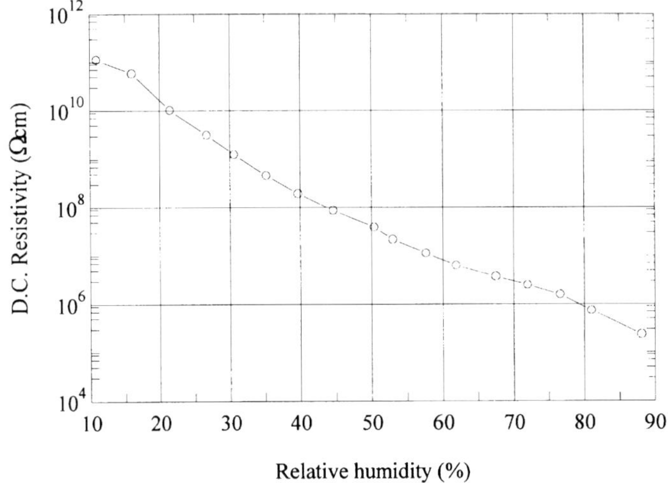

Figure 3 D.C. resistivity vs. humidity of pellet sample at 293 K.

The resistivity of the sample varied from $\sim 10^{11}$ Ωcm to $\sim 10^{5}$ Ωcm as the ambient humidity was changed from 10 %R_H to 90 %R_H at 293 K. The presently accepted model for this change of resistivity is that as the humidity is increased, water is adsorbed onto the surface of the ceramic grains, providing a path for conduction via protonic means [3].

The a.c. impedance spectra shown in figure 4 below support this model of conduction by ions in a water electrolyte.

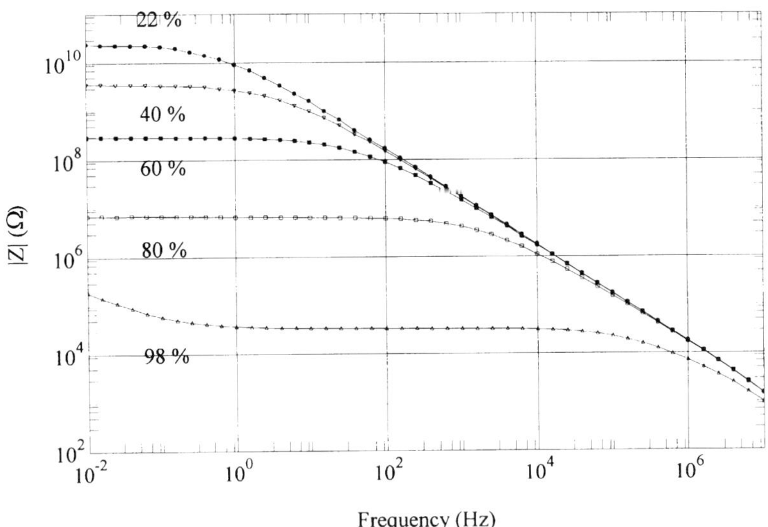

Figure 4 Change of the impedance modulus with frequency for various levels of humidity at 298 K.

The impedance characteristic at $R_H = 98$ % showed three distinct regions at high, medium and low frequency. The high frequency roll-off as |Z| becomes inversely proportional to frequency indicates capacitative dominance in the system, due to the physical geometry of the sample, cables and the input impedance of the measuring system. The almost flat response of the impedance with frequency in the middle region of figure 4 is due to the major conduction process, which in this system is believed to occur in the thin layers of water that form on the surface of the ceramic grains, with the bulk of the charge transfer occurring by protonic means. The shape of the curve at low frequencies is characteristic of a diffusive process as charge is transferred from the gold electrodes to the water in a thin interfacial layer formed on the electrodes. This explanation is further confirmed by the Cole-Cole plot of the complex impedance, figure 5.

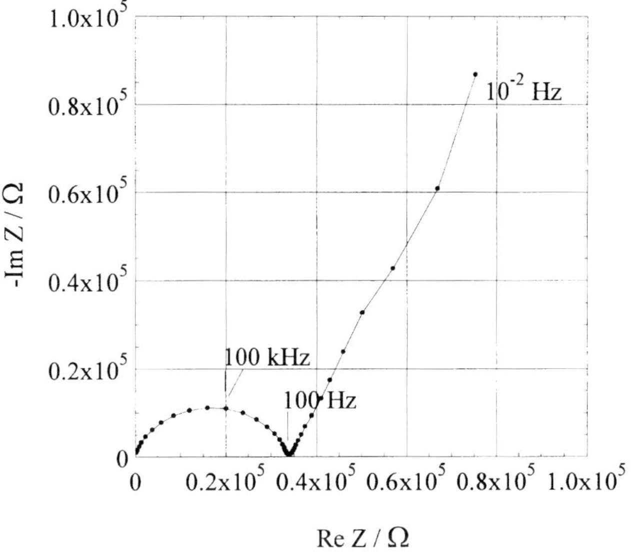

Figure 5 Cole-Cole plot of impedance at 98 %R_H

Two major features are visible on the Cole-Cole plot, a depressed semicircle, and an inclined straight line. The depressed semicircle corresponds to the flat part of the frequency response of figure 4, where charge transport by protonic means is occuring in the water layer. The inclined straight line is characteristic of diffusive charge transport across a thin interfacial layer connecting the metallic contacts to the water, which becomes the dominant process at low frequencies.

CONCLUSION

In summary, we have investigated the temperature of formation, and the electrical properties with respect to humidity, of the spinel ceramic material nickel germanate. Monophase material may be produced by firing an intimately mixed powder consisting of NiO and GeO_2 in the respective molar ratio 2:1, at 1200 °C for 12 hours.

The resistivity, and consequently the impedance, of pellet samples was noted to vary by several orders of magnitude as the humidity was changed from ~ 10 %R_H to ~ 98 %R_H, as the thickness of the adsorbed water layer coating the ceramic grains increased with increasing humidity. The a.c. impedance characteristics were found to be consistent with conduction via an electrochemical mechanism.

ACKNOWLEDGEMENTS

The authors would like to thank Elmwood Sensors UK Ltd. for their support. M. J. Hogan acknowledges the support of EPSRC.

REFERENCES

1. G. GUSMANO, G. MONTESPERELLI, P. NUNZIANTE and E. TRAVERSA: 'The electrical behaviour of $MgAl_2O_4$ pellets as a function of relative humidity', *Mater. Eng.*, 1992, **3**, 417-434.
2. S. S. PINGALE, S. F. PATIL, M. P. VINOD, G. PATHAK and K. VIJAYAMOHANAN: 'Mechanism of humidity sensing of Ti-doped $MgCr_2O_4$ ceramics', *Mater. Chem. Phys.*, 1996, **46**, 72-76.
3. E. TRAVERSA: 'Ceramic sensors for humidity detection: the state-of-the-art and future developments', *Sensors and Actuators B*, 1995, **23**, 135-156.
4. T. HASHEMI, A. W. BRINKMAN and M. J. WILSON: 'Preparation, sintering and electrical behaviour of cobalt, nickel and zinc germanate', *J. Mater. Sci.*, 1993, **28**, 2084-2088.
5. A. NAVROTSKY and O. J. KLEPPA: 'Thermodynamics of formation of simple spinels', *J. Inorg. Nucl. Chem*, 1968, **30**, 479-498.
6. A. NAVROTSKY: 'Thermodynamics of formation of the silicates and germanates of some divalent transition metals and of magnesium', *J. Inorg. Nucl. Chem.*, 1971, **33**, 4035-4050.
7. A. NAVROTSKy: 'Thermodynamic relations among olivine, spinel, and phenacite structures in silicates and germanates: II. The systems NiO-ZnO-GeO_2 and CoO-ZnO-GeO_2', *J. Solid State Chem.*, 1973, **6**, 42-47.
8. J. PREUDHOMME: 'Correlations entre spectre infrarouge et cristallochimie des spinelles', *Ann. Chim.*, 1974, **9**, 31-41.
9. M. LENGLET and C. K. JORGENSEN: 'Optical spectra of Ni(II) in Ni_2GeO_4 and germanate spinels', *Chemical Physics Letters*, 1991, **185** (1,2), 111-116.
10. K. HIROTA, T. INOUE, N. MOCHIDA and A. OHTSUKA: 'Study of germanium spinels (Part 3)', *J. Ceram. Soc. Japan*, 1990, **98** (9), 976-986.
11. J.C.P.D.S. Data card no. 10-266 (1960).
12. A. KOLLER: 'Structure and Properties of Ceramics', Elsevier Science, Amsterdam, 1994.

SOLUTION DEPOSITION AND ELECTRICAL CHARACTERISATION OF THIN PZT FILMS ON Pt/Ti/SiO_2/Si SUBSTRATES

M.H.M. ZAI, E.M. YEATMAN
Semiconductor and Optical Devices Section, Dept of Electrical and Electronic Engineering, Imperial College of Science, Technology and Medicine,
Exhibition Road, London SW7 2BT, UK
Email: m.zai@ic.ac.uk, e.yeatman@ic.ac.uk

ABSTRACT

We report on the use of a new combination of liquid metal precursors and solvent for the solution deposition of Pb(Zr, Ti)O_3 (PZT) thin films on Pt/Ti/SiO_2/Si substrates. This route uses the metal precursors lead 2-ethyl hexanoate, titanium diisopropoxide bis(acetylacetonate), and zirconium tetra n-butoxide, mixed in xylene. The PZT composition of interest is for optimum piezoelectricity near the morphotropic phase boundary (mole fraction ratio Zr/Ti = 53/47). Thin films of PZT have been made and characterised electrically. The combination of high remanent polarisation (31 $\mu C/cm^2$) with high coercive field values (148 kV/cm), and the marked dependence of permittivity with film thickness suggest that an unwanted layer is present in the PZT film. We present data supporting this view, and propose corrected permittivity values.

INTRODUCTION

There is a growing need for integrating the deposition of thin electro-ceramic films of PZT with Si micromachining, for devices like pyroelectric thermal detectors [1], surface acoustic wave (SAW) sensors [2], dynamic random-access and non-volatile ferroelectric memories (DRAM and FRAM respectively [3,4]), and the increasingly important piezoelectric microactuators [5], applied for instance to optical scanning [6] and atomic force microscopy. Our interest is in the latter type of devices.

In the fabrication of these devices, processing temperatures should be kept as low as possible, to ensure compatibility with other materials used in the device. The following techniques have been reported to allow the growth of perovskite PZT at temperatures lower than those used by conventional "powder mixing" routes: electron-beam evaporation [7], sputtering [8], chemical vapour deposition (CVD) [9], laser ablation [10], and solution deposition [11].

This work investigates the solution deposition alternative, which offers potentially the advantages of good homogeneity, ease of compositional control, low processing temperatures and compatibility with low cost equipment. Solution deposition routes can be categorised into two types: sol-gel and non sol-gel routes, depending on whether the metal precursors are, respectively, polymerised in solution, or maintain a strong likeness to the starting precursors. The main advantage of sol-gel over non sol-gel routes is the ability to prepare very viscous starting solutions, making feasible the deposition of films of one micrometer with a single layer. On the other hand, the advantage of non sol-gel routes is the much greater ease of preparing starting solutions. Here, the precursors are simply dissolved in a solvent, without the need for refluxing. In this paper, we describe a non sol-gel route, using xylene as the common solvent.

The substrate material is chosen to be $Pt/Ti/SiO_2/Si$ (150nm/50nm/500nm/wafer), widely used in the growth of Pb-based ferroelectric thin films, because Pt provides good inertness to oxygen at high temperatures, and Ti promotes adhesion between Pt and SiO_2, and prevents the formation of Pt silicides at high temperatures [12].

EXPERIMENTAL METHOD

Both Pt and Ti layers in the substrate were deposited by RF sputtering, and the SiO_2 layer was grown thermally by heating a Si(100) wafer at 1200°C for 5 hours in an oxygen-containing atmosphere.

PZT films were fabricated from a solution (1.1 M) comprising the precursors lead 2-ethyl hexanoate $[Pb(C_8H_{15}O_2)_2]$ (hereafter abbreviated LEH), titanium diisopropoxide bis(acetylacetonate) $[Ti(OC_3H_7)_2(CH_3COCHCOCH_3)_2]$ (TIAA), and zirconium tetra n-butoxide $[Zr(OC_4H_9)_4]$ (ZNBT) in xylene. This solution was stirred for about an hour in open atmosphere at room temperature. It was then filtered using a 0.2 μm syringe filter and dispensed onto a $Pt/Ti/SiO_2/Si$ substrate. The sample was spun at 4000 rpm for 40 s with a photoresist spinner, dried on a hotplate at 125°C for 1 minute, and heated in a tube furnace at 400°C for 5 minutes in air to pyrolyse the organics in the film. This sequence of spin-coating-drying-pyrolysing was iterated several times to build up thickness, the thickness per layer being around 120 nm. The sample was finally crystallised at 600°C or 700°C for half an hour in air (Fig. 1).

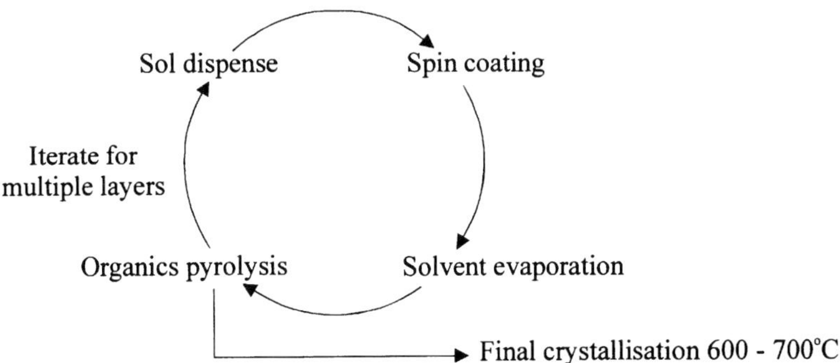

Fig. 1. Film fabrication process.

A thin adhesion layer of Cr (10 nm) was evaporated onto the PZT films, followed by a thicker layer of Au (100 nm), as the top contact for electrical characterisation. This top contact was patterned as an array of pads (with surface area around 0.5 mm^2).

The PZT films were characterised electrically using a Sawyer-Tower setup (see Fig. 2), where ferroelectric hysteresis parameters (coercive field E_c and remanent polarisation P_r) and relative permittivity ε_r were measured, respectively, at higher and lower input voltages. All these measurements were carried out at frequencies around 1kHz. Film thicknesses, necessary for the determination of ε_r and E_c, were measured by cross-sectional scanning electron microscope (SEM), or by profiling with a scanning tip a film patterned by lithography and wet etching.

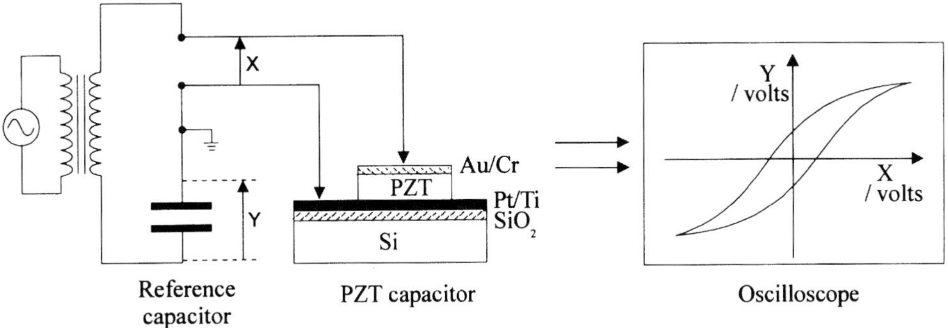

Fig. 2. Sawyer-Tower setup for electrical measurements.

RESULTS

Fig. 3 shows a typical ferroelectric hysteresis measured for a PZT film made from a solution with 20% Pb excess and a Zr/Ti ratio of 53/47. This film was crystallised at 600°C for 30 minutes and is 0.5 μm thick. The remanent polarisation P_r is around 31 μC/cm², and the coercive field E_c is around 150 kV/cm. Fig. 4 shows measurements made on a similar film, but crystallised at 700°C. The measured values in this case are E_c = 140 kV/cm and P_r = 30 μC/cm².

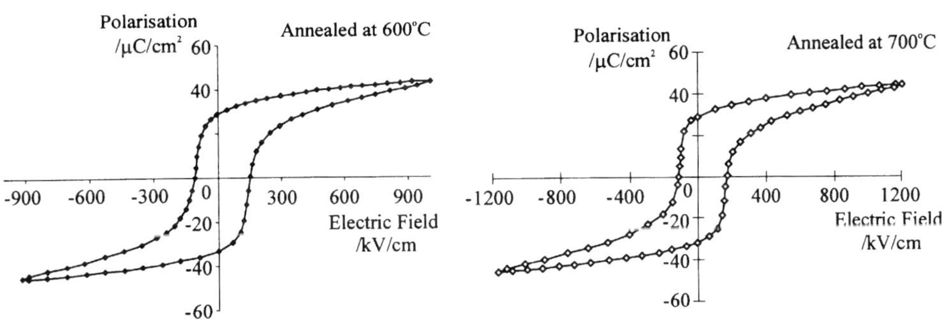

Fig. 3. Ferroelectric properties of film crystallised at 600°C.

Fig. 4. Ferroelectric properties of film crystallised at 700°C.

Additional amounts of Pb were introduced in the starting solutions to compensate for the well-known problem of Pb loss occurring as a result of the formation of volatile PbO. Figs. 5 and 6 show the relative permittivity plotted against Pb excess for crystallisation at temperatures 600°C and 700°C, for half an hour. For each value of Pb excess, data points are presented for different contact pads on the same sample. These results suggest that an excess of Pb in the starting solutions improves the quality of the PZT films.

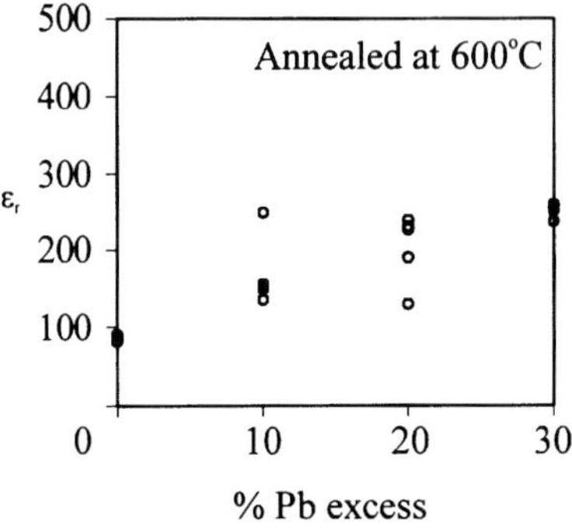

Fig. 5. Permittivity vs. Pb excess (Crystallised at 600°C)

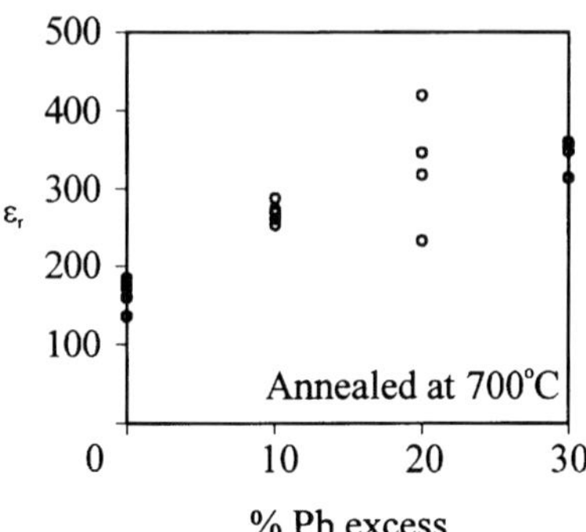

Fig. 6. Permittivity vs. Pb excess (Crystallised at 700°C)

Using a PZT solution with Zr/Ti = 50/50, the relative permittivity ε_r of the films was measured as a function of thickness (see Figs. 7 and 8).

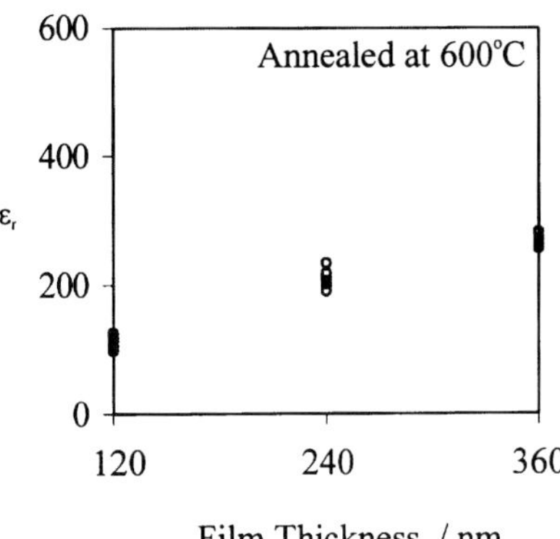

Fig. 7. Permittivity vs. Film thickness (Crystallised at 600°C)

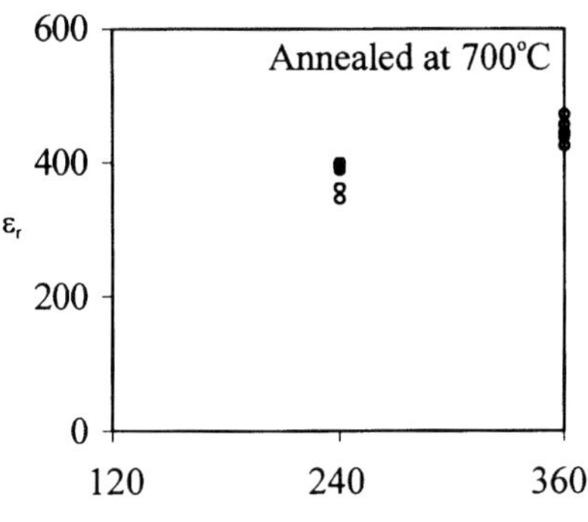

Fig. 8. Permittivity vs. Film thickness (Crystallised at 700°C)

DISCUSSION

Table I shows a comparison of ferroelectric hysteresis measurements for this work with published results.

Table I - Comparison with published results.

Ref.	Precursors				Starting Composition		Process		Electrical properties	
					Zr / Ti	Pb excess	Thickness	Substrate	P_r	E_c
	Pb	Ti	Zr	Solvent		/ mol %	/ µm		/ µC/cm^2	/ kV/cm
[13]	LEH	TiPT	ZNBT	Hexane	53 / 47	?	0.41	Pt/SiO_2/Si	15	35
[14]	LAT	TIAA	ZRAA	PD	53 / 47	10	0.5	Pt/Si(100)	34	45
[15]	LAT	TiPT	ZrPT	EG	52 / 48	?	0.3	Pt/Ti/SiO_2/Si(111)	15	60
[16]	LEH	TiPT	ZrPT	n-PrOH	54 / 46	5	0.62	Pt/Si(100)	25	50
Current work	LEH	TIAA	ZNBT	xylene	53 / 47	20	0.5	Pt/Ti/SiO_2/Si(100)	31	148

LEH = lead 2-ethylhexanoate; LAT = lead acetate trihydrate;
TIAA = titanium diisopropoxide bis(acetylacetonate); TiPT = titanium tetra-isoproproxide
ZNBT = zirconium tetra n-butoxide; ZrPT = zirconium tetra n-propoxide
EG = ethylene glycol; PD = 1,3-propanediol

The combination of good remanent polarisation with poor coercive field values (Figs. 3 and 4), and the marked dependence of permittivity on thickness (Figs. 7 and 8), support the supposition that a low dielectric constant layer could be present in the PZT capacitor. This unwanted phase could have arisen from either a poorly deposited top contact or a chemical reaction between bottom electrode and PZT film.

We model the film as a reactance divider, or two capacitors in series, one representing the ferroelectric PZT phase and another representing the unwanted layer. Here, voltage across a capacitor is inversely proportional to its capacitance, or proportional to d/ε_r, where d is the thickness of the dielectric in the capacitor. *As a result, a large permittivity thin film yields a small voltage, which is difficult to measure when a lower permittivity dielectric phase, introducing a large "noise" signal, is present, even in very small quantities.*

To estimate the properties of the PZT phase alone, the contribution of the unwanted layer to the measurements must be cancelled. For simplicity, we assume that this layer is much thinner than the PZT, and that its properties are similar for PZT films of different thicknesses. We subtract the inverse of the capacitances of PZT films of different thicknesses to extract the capacitance of the pure PZT phase alone (Fig. 9).

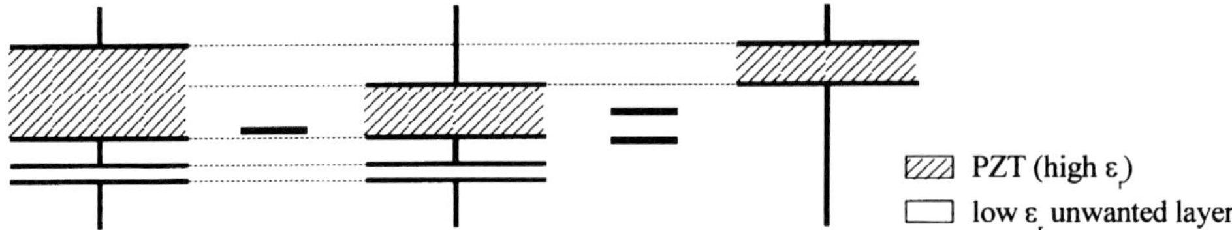

Fig. 9. Reducing the contribution of the unwanted layer to the measurements.

Permittivity can be extracted by measuring the slope of the linear curve fitting, since,

$$C = \varepsilon_0 \varepsilon_r A / d \tag{1}$$

where C, ε_0, A, and d are the measured capacitance, the free space permittivity, the contact pad area, and the film thickness respectively. As a result,

$$1/\varepsilon_r = (\varepsilon_0 A/C) / d \tag{2}$$

By replotting our data, we obtain Figs. 10 and 11.

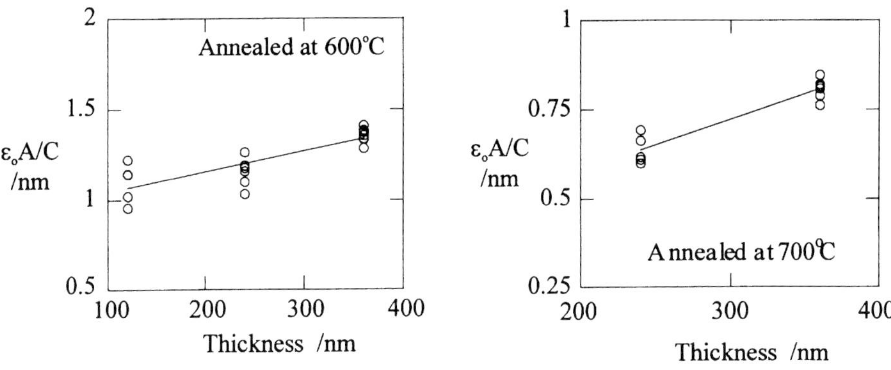

Fig. 10. Extracting the permittivity of PZT. (Crystallised at 600°C)

Fig. 11. Extracting the permittivity of PZT. (Crystallised at 700°C)

The estimated values are around ε_r =900 and ε_r = 700 for films annealed at 600°C and 700°C respectively (Table II), with a precision of ±100 based on scatter in the data, but they and the model used must be confirmed by physically eliminating the unwanted layer, which we are currently aiming to do.

Table II - Estimated permittivity values

		Mean permittivity of PZT			
		Measured			Estimated (+/-100)
Film Thickness / Angstroms		1200	2400	3600	
Annealing Temperature	600°C	110	207	265	900
	700°C	NA	378	447	700

NA : Not available

CONCLUSIONS

We have fabricated successfully PZT thin films on $Pt/Ti/SiO_2/Si$ substrates using a solution route with a new combination of metal precursors in xylene. These films have been characterised electrically using a Sawyer-Tower circuit. The remanent polarisation measurements were good (~30 $\mu C/cm^2$), but coercive fields (~150 kV/cm) and relative permittivity (< 410) were poor. We suggested that this situation could be caused by the presence of a low permittivity layer; and assuming that this unwanted layer is scarce, the permittivity of PZT is estimated to be as high as 900.

ACKNOWLEDGEMENTS

We thank Dr. P. Judge for the deposition of Pt and Ti, and Prof. R.R.A. Syms, Dr. M.M. Ahmad, and Dr. S. Wright for helpful discussions. We are also grateful for partial financial support by the European Commission (ACTS project AC047).

REFERENCES

[1] R. KOHLER, N. NEUMANN, N. HESS, R. BRUCHHAUS, W. WERSING, and M. SIMON: 'Pyroelectric Devices Based on Sputtered PZT Thin Films', *Ferroelectrics*, 1997, **201** (1-4), 83-92.

[2] W.C. SHIH and M.S. WU: 'Effect of a Buffer Layer on the Surface Acoustic Wave Characteristics of $Pb(Zr,Ti)O_3$ Film/Buffer Layer/Semiconductor Substrate Structures', *Japanese Journal of Applied Physics Part 1 -Regular Papers Short Notes & Review Papers*, 1997, **36** (1A), 203-208.

[3] J.F. SCOTT and C.A. PAZ DE ARAUJO: 'Ferroelectric Memories', *Science*, 1989, **246**, 1400-1405.

[4] R.E. JONES, JR. and SESHU B. DESU: 'Process Integration for Nonvolatile Ferroelectric Memory Fabrication', *MRS Bulletin*, 1996, 55-58.

[5] W.P. ROBBINS: 'Ferroelectric-Based Microactuators', *Integrated Ferroelectrics*, 1995, **11**, 79-190.

[6] G.H. YI, Z. WU, and M. SAYER: 'Preparation of $Pb(Zr,Ti)O_3$ Thin Films by Sol-Gel Processing - Electrical, Optical, and Electro-optic Properties', *Journal of Applied Physics*, 1988, **64** (5), 2717-2724.

[7] B.E. PARK, I. SAKAI, E. TOKUMITSU, and H. ISHIWARA: 'Hysteresis Characteristics of Vacuum-Evaporated Ferroelectric $PbZr_{0.4}Ti_{0.6}O_3$ Films on Si(111) Substrates Using CeO_2 Buffer Layers', *Applied Surface Science*, 1997, **117**, 423-428.

[8] T.S. KIM, D.J. KIM, J.K. LEE, and H.J. JUNG: 'Fabrication of Excess PbO-Doped $Pb(Zr_{0.52}Ti_{0.48})O_3$ Thin Films Using Radio Frequency Magnetron Sputtering Method', *Journal of Vacuum Science & Technology, A-Vacuum Surfaces and Films*, 1997, **15** (6), 2831-2835.

[9] H. YAMAZAKI, T. TSUYAMA, I. KOBAYASHI, and Y. SUGIMORI: 'Preparation of $Pb(Zr, Ti)O_3$ Thin-Films Using All Dipivaloylmethane Source Materials By Metalorganic Chemical Vapor-Deposition', *Japanese Journal of Applied Physics, Part 1-Regular Papers Short Notes & Review Papers*, 1992, **31** (9B), 2995-2997.

[10] J. LAPPALAINEN, J. FRANTTI, H. MOILANEN, and S. LEPPAVUORI: "Excimer Laser Ablation of PZT Thin Films on Silicon Cantilever Beams," *Sensors and Actuators*, 1995, **A 46-47**, 104-109.

[11] K. AMANUMA, T. MORI, , T. HASE, and T. SAKUMA: "Ferroelectric Properties of Sol-Gel Derived $Pb(Zr, Ti)O_3$ Thin Films," *Jpn. J. Appl. Phys.*, 1993, **32**, 4150-4153.

[12] J.O. OLOWOLAFE, R.E. JONES, JR., A.C. CAMPBELL, R.I HEGDE, C.J. MOGAB, and R.B. GREGORY: 'Effects of Anneal Ambients and Pt Thickness on Pt/Ti and Pt/Ti/TiN Interfacial Reactions', *J. Appl. Phys.*, 1993, **73** (4), 1764-1772.

[13] B.A. TUTTLE, T.J. HEADLEY, B.C. BUNKER, R.W. SCHWARTZ, T.J. ZENDER, C.L. HERNANDEZ, D.C. GOODNOW, R.J. TISSOT, and J. MICHAEL: 'Microstructural Evolution of $Pb(Zr, Ti)O_3$ Thin Films Prepared by Hybrid Metallo-Organic Decomposition', *J. Mater. Res.*, 1992, **7** (7), 1876-1882.

[14] Y.-L. TU, M.L. CALZADA, N.J. PHILLIPS, and S.J. MILNE: 'Synthesis and Electrical Characterisation of Thin Films of PT and PZT Made from a Diol-Based Sol-Gel Route', *J. Am. Ceram. Soc.*, 1996, **79** (2), 441-448.

[15] C. LIVAGE, A. SAFARI, and L.C. KLEIN: 'Glycol-Based Sol-Gel Process for the Fabrication of Ferroelectric PZT Thin Films', *Journal of Sol-Gel Science and Technology*, 1994, **2**, 605-609.

[16] C.-C. HSUEH and M.L. MECARTNEY: 'Microstructural Development and Electrical Properties of Sol-Gel Prepared Lead Zirconate-Titanate Thin Films', *J. Mater. Res.*, 1991, **6** (10), 2208-2217.

PROCESSING AND ELECTRICAL PROPERTIES OF LEAD ZIRCONATE TITANATE (PZT) CERAMICS FROM HYDROTHERMALLY SYNTHESISED POWDERS

B. SU*, D.H. PEARCE, T.W. BUTTON AND C.B. PONTON
IRC in Materials for High Performance Applications & School of Metallurgy and Materials, The University of Birmingham, Edgbaston, Birmingham B15 2TT, UK
Email: t.w.button@bham.ac.uk

ABSTRACT

The processing and piezoelectric properties of the PZT ceramics produced from hydrothermally synthesised powders have been investigated and compared with ceramics produced from mixed-oxide powders. The compaction and sintering behaviour of the PZT powders processed via colloidal processing and uniaxial dry powder pressing routes have been examined. The results show that the sintering temperature for hydrothermally synthesised powders was about 100 to 200°C lower than that for the mixed-oxide powders. Utilisation of colloidal processing routes decreased the sintering temperature further. The dielectric constant was found to be largely dependent on the physical state (e.g. density) of the ceramics whereas the piezoelectric properties seemed more sensitive to the composition and nature of the dopants. The correlation of the electrical properties with the composition and microstructure changes during processing is discussed.

INTRODUCTION

Lead zirconate titanate (PZT) ceramic is a widely used ferroelectric material in piezoelectric devices. The conventional way to fabricate such ceramics is via the mixed-oxide powder processing route. There are several disadvantages in connection with this processing route, such as chemical inhomogeneity due to the different diffusion rates of each component during solid-state reaction, impurities possibly introduced during milling, lead loss during calcining and sintering, and high energy consumption because of the repeated wet mixing/drying process. Alternative methods have been investigated in recent years. Among them, hydrothermal synthesis is particularly promising for making multi-component mixed-oxide systems such as PZT powders. We have reported that the morphology and particle size distribution of the perovskite PZT powders can be controlled during hydrothermal synthesis and submicron-metre PZT powders with various dopants have been obtained[1]. In this paper, we examine the compaction and sintering behaviours of the hydrothermal PZT powders by employing a colloidal processing route. Comparison is made with a mixed-oxide PZT powder and with uniaxial dry powder pressing. The dielectric and piezoelectric properties of the resultant PZT ceramics are discussed in accordance with their microstructure and composition changes during processing.

EXPERIMENTAL

Five hydrothermal PZT powders have been used and compared with two mixed-oxide powders. The specifications of each of the PZT powders are listed in Table I. A non-aqueous polymer binder was used in the dry powder pressing. Green disc compacts were pressed in a one-action die with diameter of 13 mm at a pressure of 100 MPa.

* Current address: Department of Materials, Imperial College, Prince Consort road, London SW7 2BP, UK

Table I. Specifications of the PZT powders used in uniaxial dry powder pressing.

PZT Powder*	Synthesis Route	Composition and Synthesis Conditions (The basic composition is $Pb(Zr_{0.52}Ti_{0.48})O_3$)
HS-36	hydrothermal	undoped, 4 M KOH, 200°C/0.5 hour, one-step feedstock
HT-29	hydrothermal	undoped, 0.4 M KOH, 300°C/2 hours, two-step feedstock
HT-34	hydrothermal	1 mol % Nb, 0.4 M KOH, 300°C/2 hours, two-step feedstock
HT-35	hydrothermal	1 mol % La, 0.4 M KOH, 300°C/2 hours, two-step feedstock
HT-37	hydrothermal	0.6 mol % Nb, 0.4 M KOH, 300°C/2 hours, two-step feedstock
PZT-5A	mixed-oxide	unknown doping, mixing, jet milling, calcination
MO-PZT	mixed-oxide	2 mol % Nb, mixing, ball milling, calcination

* PZT-5A and MO-PZT are two mixed-oxide PZT powders from Morgan Matroc (Verniton, Southampton, UK) and self-made respectively. For details about the preparation of the hydrothermal PZT powders see Ref. 1.

Viscous processed green compacts were prepared by mixing the PZT powders with an aqueous polymer system in a water-cooled twin-roll milling machine (Winkworth Machinery Ltd, UK) for about 30 minutes. The resultant green sheet was then placed between two steel plates in a hydraulic press at a pressure of about 1 MPa for 24 hours for de-airing. Disc samples were subsequently cut from the sheet using a cork borer and then dried between two flat plaster plates at 40°C for 12 hours. Debinding was carried out in an oven using a heating rate of 1°C/min and a holding temperature of 600°C for 2 hours before sintering.

Sintering was carried out in a furnace under a PZT buffer atmosphere using a heating rate of 1°C/min to different maximum temperatures from 750°C to 1250°C for holding times from 5 minutes to 2 hours. The sintering behaviour in air was monitored via a differential dilatometer (Netzsch, Geratebau GmBH, Germany), using a sapphire reference and a heating rate of 10°C/min. The sintered density of specimens was determined using the Archimedes' method. Silver paste electrodes were spray-painted on both sides of the specimens and fired in air at 600°C for 10 min. The specimens were poled under a DC field of 2.5 kV/mm in a mineral oil bath at 120°C for 10 min. All the measurements were performed after the poling treatment, using an impedance analyser (4194A, Hewlett Packard, USA). The relative permittivity ε_r was measured at a frequency of 1 kHz. The electromechanical coupling coefficient K_p and the mechanical quality factor Q_m were measured by a resonance-antiresonance method[2].

RESULTS AND DISCUSSION

Compaction Behaviour

Viscous processing or VP involves the use of polymer solutions in a particular way in the processing of ceramic powders. It enables the powder agglomerates to be broken down more easily by using the viscous polymer solution to transfer significant stresses to the powder agglomerates, and also confers a number of benefits to the product such as better sintering behaviour, greater homogeneity and higher strength [3].

In this work it was found that the necessary polymer content is strongly dependent on the powder characteristics. The amount of polymer required to convert the hydrothermal PZT powder (HT-37) into a workable dough is almost twice that required for the mixed-oxide PZT powder (PZT-5A). This is because of the larger surface areas of the hydrothermal PZT powder particles. However, no significant difference between their green densities was evident after polymer burnt-out. It is noteworthy that the green compacts processed via colloidal processing (VP), exhibit higher green densities than the dry pressed compacts, despite the applied pressure being lower (see Table II). Particles in the colloidal state are much less resistant to rearranging and dispersion prevents further agglomeration, so a more uniform microstructure is achieved. This is attributed mainly to the modification of the interparticle forces by the added dispersant.

Table II. Green density of the PZT ceramics produced by dry powder pressing and VP.

Processing	Dry Pressing	VP
Green density of PZT-5A (g/cm^3)	4.59	4.80
Relative density of PZT-5A (%)*	58.8	61.5
Green density of HT-37 (g/cm^3)	3.81	4.07
Relative density of HT-37 (%)*	58.6	62.6

* The relative density is defined as the ratio of green density to powder density. The true powder density as measured by the density bottle method for PZT-5A and HT-37 was 7.8 and 6.5 g/cm^3, respectively.

Sintering Behaviour

Fig. 1 shows the variation of sintered density with sintering temperature for both hydrothermal and mixed-oxide PZT powders processed via dry pressing and VP routes respectively after sintering for a fixed time of 2 hours. The benefits of using colloidal processing (VP) and smaller sized hydrothermal powders are clearly evident. The optimum sintering temperature (i.e. the lowest temperature at which the maximum density is achieved) for the hydrothermal HT-37 powder is significantly lower (about 250°C) than that for the mixed-oxide PZT-5A powder. Also, the sintering temperature for the VP-processed ceramic is about 100-150°C lower than that for the dry-pressed ceramic. For HT-37 with a finer particle size and narrower particle size distribution[1], the sintered density increases abruptly in the sintering temperature range of 750 to 900°C, whereas for PZT-5A powder, it increases gradually and reaches the maximum at significantly higher temperatures than that for HT-37.

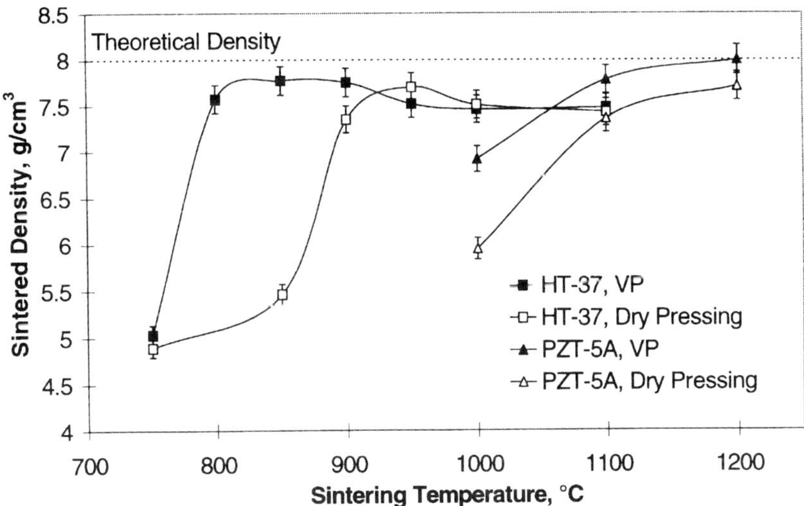

Figure 1. Sintered density versus sintering temperature (for a fixed time of 2 hours) for both hydrothermal (HT-37) and mixed-oxide (PZT-5A) ceramics, produced by VP and by dry pressing.

Dielectric and Piezoelectric Properties

Fig. 2 (a) shows the dielectric and piezoelectric properties, as represented by ε_r, the relative permittivity, and K_p, the piezoelectric coupling coefficient, respectively, for an undoped hydrothermal PZT (HT-29) and a doped mixed-oxide (PZT-5A) ceramic as a function of the sintering temperature, at a short holding time of 5 minutes. Their corresponding sintered density and grain size changes are shown in Fig. 2 (b). It can be seen that ε_r increases with the density, whereas K_p seems to be more related to the grain size of the PZT ceramic. For the undoped HT-29 ceramic, a maximum ε_r of about 1150 has been reached at 850°C, followed by a slight decrease at higher temperatures due to the minor decrease of the density owing to the lead loss during sintering. The K_p, however, increases dramatically only when the grain size is above 1.8 μm. Since the K_p is more related to the mobility of the domain wall in the PZT ceramics, the increase of grain size results in more domains and, therefore, possibly more mobile walls[4]. In ceramics with a larger grain size, multi-domain effects accompanied by hysteresis are more likely to take place. Reduction in grain size leads to single-domain effects and, finally, at sufficiently small sizes, the number of domains is expected to diminish[5]. Mishra and Pandey[6] reported that the formation of ferroelectric domains is not energetically favourable below a critical grain size. However, this critical size is dependent on the material and its composition. For the doped PZT-5A ceramic, the increase in K_p starts at a lower grain size (i.e. 0.5 μm for the doped PZT-5A versus 1.8 μm for the undoped HT-29 ceramic) and K_p increases dramatically to 0.65 as the grain size increases to 2.2 μm. This implies that the doping ions increase the domain wall mobility[7].
The permittivity is a measurement of the polarizability which relates an induced dipole moment in an atom, ion, molecule etc., to the internal field inducing the dipole moment. Obviously, the permittivity of a ferroelectric ceramic depends on both its intrinsic structure and its physical state. For a PZT ceramic with small inclusions of gas (porosity < 10 vol %), the relative permittivity obeys a rule of mixtures[8]. The relation is as follows:

$$\varepsilon_r = \varepsilon_r' \,(1\text{-}3\,P/2) \qquad (1)$$

where ε_r and ε_r' are the relative permittivities of the porous and fully dense ceramics respectively, and P is the porosity. Assuming the measured true density and the theoretical density of the PZT ceramic is ρ and ρ_{th}, respectively, then

$$\varepsilon_r = \varepsilon_r' \,[1\text{-}3\,(1\text{-}\,\rho/\rho_{th})/2] \qquad (2)$$

Fig. 2 (a) also shows the comparison of the calculated ε_r for the hydrothermal (HT-29) and mixed-oxide (PZT-5A) PZT ceramics during the sintering process. It can be seen that the calculated and measured ε_r is only in agreement for the undoped HT-29. A large discrepancy has been found for the doped PZT ceramics (PZT-5A), which may be attributed to its relatively low density and thus high porosity at lower sintering temperatures.
The electrical properties of sintered samples of both the mixed-oxide and hydrothermal PZT ceramics sintered at different conditions are given in Table III. Comparing the electrical properties of the doped and undoped PZT ceramics in Table III, it can be seen that the K_p and ε_r values for the Nb or La doped PZT ceramics, whether from a mixed-oxide (e.g. MO-PZT) or hydrothermal synthesis (e.g. HT-37, HT-34, HT-35) route, are generally higher than those of undoped PZT ceramics (e.g. HT-29, HS-36). In addition, the K_p and ε_r values seem to increase with increasing dopant content (HT-37, HT-34, MO-PZT), whereas their Q_m is generally lower than that of the undoped PZT ceramics. It can also be seen that the

hydrothermal PZT ceramic (HS-36) with 3 mol % K residue has much higher Q_m than other PZT ceramics.

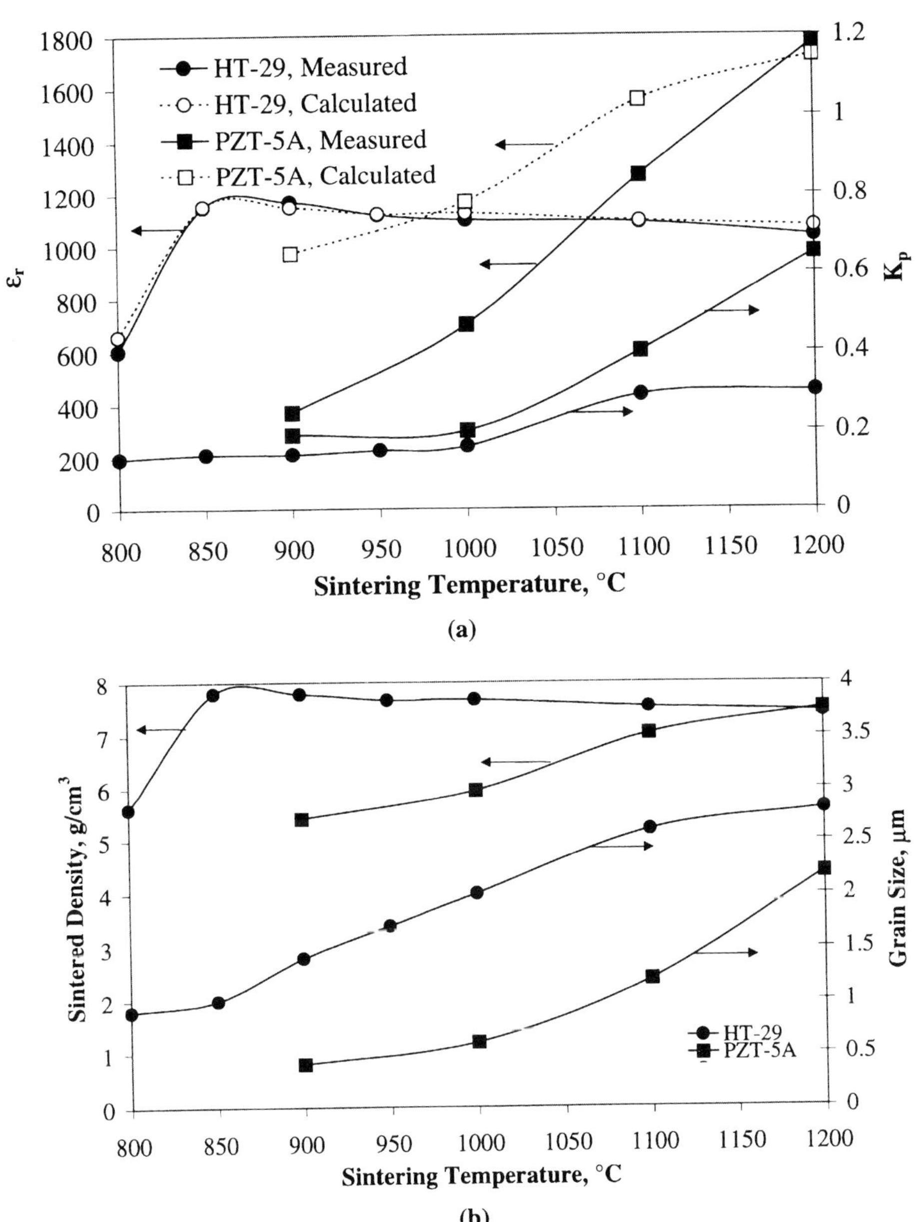

Figure 2. Comparison of the electrical properties and microstructure changes of the undoped hydrothermal (HT-29) ceramic and the doped mixed-oxide (PZT-5A) ceramic as a function of the sintering temperature for a fixed sintering time of 5 minutes.

When PZT ceramics are doped with "soft dopants" such as Nb or La in this study, the doping ions will either occupy the A-sites or B-sites to replace Pb^{2+} or Zr^{4+} and Ti^{4+} and create more Pb vacancies owing to their different valences. The Pb vacancies make the transfer of atoms easier than in a perfect lattice, thus increasing the domain wall mobility[7]. Therefore, both K_p and ε_r are substantially higher, whereas the Q_m is lower in comparison with the undoped PZT ceramics.

Table III. Electrical properties of sintered PZT ceramics.

Sample	Composition	Sintering Conditions	Density(g/cm^3)	K_p	ε_r	Q_m
HT-29	$Pb(Zr_{0.52}Ti_{0.48})O_3$	1200°C/2 hours	7.49	0.42	989	147
HT-37	$Pb(Zr_{0.52}Ti_{0.48})O_3$ + 0.6 mol % Nb	1000°C/2 hours	7.51	0.55	1490	53
HT-34	$Pb(Zr_{0.52}Ti_{0.48})O_3$ + 1.0 mol % Nb	1100°C/2 hours	7.80	0.62	1610	48
HT-35	$Pb(Zr_{0.52}Ti_{0.48})O_3$ + 1.0 mol % La	1100°C/2 hours	7.43	0.60	1780	51
HS-36	$Pb(Zr_{0.52}Ti_{0.48})O_3$ + 3.0 mol % K	1200°C/2 hours	7.20	0.46	995	546
MO-PZT	$Pb(Zr_{0.52}Ti_{0.48})O_3$ + 2.0 mol % Nb	1200°C/2 hours	7.70	0.63	1700	46
PZT-5A	doped, unknown	1200°C/2 hours	7.71	0.65	1790	42

The presence of potassium in the hydrothermal PZT ceramic (HS-36) is caused by unintentional contamination during the hydrothermal synthesis where KOH is used as a mineraliser. EDS analysis shows that a small amount of potassium is present within the hydrothermal PZT ceramics (about 1-3 mol %) depending on the feedstock preparation method and mineraliser concentration used, especially for PZT powders from one-step derived feedstock where a high mineraliser concentration is necessary for the perovskite PZT formation[1]. Potassium ions are known as hard dopants in PZT, which occupy the A sites in the perovskite structure, generating oxygen vacancies in the perovskite. Sintering the ceramic in an atmosphere of oxygen cannot eliminate these oxygen vacancies. This may explain the experimental observation that almost all the undoped PZT ceramics are dark grey in colour as reported in other ferroelectric ceramics with oxygen vacancies[8].

It has also been confirmed that PZT ceramics possess p-type conductivity, i.e. act as hole-type semiconductors[9]. In a PZT ceramic with hard doping ions, the number density of space charges (i.e. both centres of negative charge and hole carriers) increases dramatically. These space charges cause an internal electrical field inside the PZT grains, which may inhibit the movement of ferroelectric domain walls, resulting in the increase of Q_m[10].

CONCLUSIONS

Hydrothermal synthesis in conjunction with colloidal processing has shown to be beneficial to the reduction of sintering temperatures of PZT ceramics. The permittivity was found to be dependent on the sintered density but piezoelectric properties seems more dependent on the composition and homogeneity of the PZT ceramics.

ACKNOWLEDGEMENTS

The authors acknowledge Morgan Matroc for providing PZT-5A powder. One of the authors (B. Su) would like to thank the financial support from the ORS Award and the School of Metallurgy and Materials, The University of Birmingham.

REFERENCES

1. B. SU, T.W. BUTTON and C.B. PONTON: 'Control of the Particle Size and Morphology of Hydrothermally Synthesised Lead Zirconate Titanate (PZT) Powders', submitted to *J. Am. Ceram. Soc.*, 1997.
2. IRE Standards on Piezoelectric Crystals: 'Measurements of Piezoelectric Ceramics', *Proc. IRE,* 1961, **49,** 1162.
3. N. McN, ALFORD, J. D. BIRCHALL and K. KENDALL: 'High-Strength Ceramics Through Colloidal Control to Remove Defects', *Nature*, 1987, **330**, [6143] 51-53.
4. H. T. MARTRENA and J. C. BURFOOT: 'Grain Size Effects on Properties of Some Ferroelectric Ceramics', *J. Phys. C: Solid State Phys.*, 1974, **7**, 3182-3192.
5. R. E. NEWNHAM, K. R. UDAGAKUMAR and S. TROLIER-MCKINSTRY: 'Size Effect in Ferroelectric Thin Films'; pp. 379-393 in Chemical Processing of Advanced Materials, Ed. By L. L. Hench and J. K. West, John Wiley & Sons Inc., New York, 1992.
6. S. K. MISHRA and D. PANDEY: 'Effect of Particle Size on the Ferroelectric Behaviour of Tetragonal and Rhombohedral $Pb(Zr_xTi_{1-x})O_3$ Ceramics and Powders', *J. Phys.: Condens. Matter*, 1995, **7,** 9287-9303.
7. Y. XU: pp. 101-210 in *Ferroelectric Materials and Their Applications*, North Holland, 1991.
8. A. J. MOULSON and J. M. HERBERT: pp. 265-276 in *Electroceramics: Materials, Properties and Applications*, Chapman & Hall, 1991.
9. R. GERSON: 'Variation in Ferroelectric Characteristics of Lead Zirconate Titanate Ceramics Due to Minor Chemical Modifications', *J. Appl. Phys.*, 1960, **31** [1] 188-194.
10. S. TAKAHASHI: 'Internal Bias Field Effects in Lead Zirconate-Titanate Ceramics Doped with Multiple Impurities', *Jpn. J. Appl. Phys.*, 1981, **20** [1] 95-101.

PREPARATION AND CHARACTERISATION OF MOLYBDENUM-DOPED PZT CERAMICS FOR AN INVESTIGATION OF PIEZOELECTRIC ACTIVITY

L. BLACKWELL*, D. HIND** and N. W. THOMAS***
*Dept of Materials, University of Leeds, Leeds, LS2 9JT
Email mselb@leeds.ac.uk
** Dept of Materials, University of Leeds, Leeds, LS2 9JT
Email cerdah@leeds.ac.uk
* now at Watts Blake Bearne & Co PLC, Newton Abbot, Devon, TQ12 4PS
Email nthomas@wbb.co.uk

ABSTRACT

Lead Zirconium Titanate (PZT) materials doped with Mo^{6+} have been synthesised through a chemical route. An investigation of relationships between the temperature of the Curie point and Mo^{6+} doping level shows that values are influenced by the sintering atmosphere. A study of poling behaviour demonstrates improvements by 1at% Mo^{6+} substitutions in PZT, which can be related to the introduction of A-site vacancies increasing domain wall mobility. The influence of an oxygen atmosphere during sintering, is described in terms of a 'softening' effect in undoped PZT compared with a 'hardening' effect in Mo^{6+} doped PZT.

INTRODUCTION

Lead zirconate titanate (PZT) has been the focus of a significant amount of work since the superior piezoelectric properties of this material were first discovered. Investigations by many authors have concerned both the preparation of PZT and its ferroelectric and piezoelectric properties[1-3].

A wide range of dopants have been employed in such studies. The variations in the conduction, dielectric and piezoelectric properties of PZT observed with donor and acceptor doping have been explained in terms of domain wall mobility and Curie point. Oxygen vacancies are described as being detrimental to domain wall motion, limiting their orientation with poling fields. In comparison A-site vacancies are linked to increased domain wall motion and therefore improved poling behaviour[1,4-7]. In some reported work, it is however, unclear as to the intended occupancy site of the dopant and there is a lack of reference to the effect of ionic size as a factor.

The possibility of compositional inhomogeneity is clearly an issue in any examination of the properties of PZT and is particularly important where 'traditional' mixed oxide routes have been used[8]. Other techniques have been used as a means of avoiding this problem[1,9] and a range of these are now available for the preparation of ceramic powders[8,10,11]. These offer improved homogeneity and compositional control compared with mixed oxide routes[8] and, in particular, the spray drying of nitrates has been shown as a suitable technique for the formation of high purity, homogenous PZT-based materials[4,9].

This work covers the influence of Mo^{6+} on the piezoelectric behaviour of PZT. The effects of molybdenum (VI) oxide additions, via a mixed oxide route, on the resistivity, dielectric and radial coupling factor of PZT are reported elsewhere[5-7]. In the present investigation powders were prepared by a technique which involved the spray drying of aqueous nitrate precursors and sintering of PZT in an oxygen atmosphere. This is an established route to higher density ceramics as it removes insoluble nitrogen, which can limit pore shrinkage and consequently hinder densification[12-14].

EXPERIMENTAL

Precursor materials were titanium isopropoxide (99.97% Strem Chemicals Inc.), zirconyl nitrate solution (99.99% Magnesium Elektron Ltd.), lead nitrate (99.5% Alfa Products) and ammonium molybdate (99.99% Aldrich Chemical Co.). The titanium isopropoxide was hydrolysed to form hydrated titania which was dissolved in dilute nitric acid to produce a titanyl nitrate solution. Chemical analysis of the titanyl nitrate and zirconyl nitrate solutions was carried out to determine the absolute concentrations of titanium and zirconium.

Doped and undoped PZT compositions were prepared from dilute solution of lead nitrate in distilled water combined with the titanyl and zirconyl nitrate solutions. Molybdenum (VI) doped compositions were prepared by the addition of ammonium molybdate dissolved in dilute nitric acid. The resulting solution was spray dried using a Buchi 190 Mini Spray Drier with the process parameters given in table I.

Table I: Spray drying process parameters for the preparation of PZT precursor powder..

Process parameter	**Setting**
Product feed rate	10ml/minute
Drying air temperature	215°C
Outlet temperature	100 - 130°C
Compressed air flow	600 litres/hour at r.t.p.

The specific compositions prepared and the identifiers adopted are given in table II.

Table II: Doped and undoped PZT compositions prepared in this work with associated identifier.

Composition	Identifier
$PbZr_{0.52}Ti_{0.48}O_3$	PZT 52/48
$PbZr_{0.57}Ti_{0.43}O_3$	PZT 57/43
$PbZr_{0.62}Ti_{0.38}O_3$	PZT 62/38
$Pb_{0.99}Mo_{0.01}Zr_{0.52}Ti_{0.47}O_3$	PMZT 1/52/47
$Pb_{0.99}Mo_{0.01}Zr_{0.57}Ti_{0.42}O_3$	PMZT 1/57/42
$Pb_{0.99}Mo_{0.01}Zr_{0.62}Ti_{0.37}O_3$	PMZT 1/62/37

Calcination of powders in silica dishes took place at 650°C (compared with 800-900°C typically reported elsewhere for mixed oxide routes) for 4 hours. XRD analysis of the calcined powders confirmed the formation of single phase PZT. Powders were micronized and pressed (100MPa) into 15mm diameter discs. A sintering study was carried out, both in air and oxygen, at temperatures between 1125 - 1300°C, with a 'lead atmosphere' source provided using lead zirconate powder. Sample densities were measured geometrically on samples in the form of plates cut from the sintered discs, assuming a theoretical density, ρ_{theo} of 8Mg m^{-3} for all materials.

To determine the effect of Mo^{6+} addition on the tetragonal/rhombohedral phase ratio XRD analysis of samples of optimum density, sintered in both air and oxygen, was carried out. Measurements of relative permittivity versus temperature were made using a HP4194A impedance analyser, in order to determine the Curie point.

Disc shaped samples were poled with successively increasing fields in a silicone oil bath for 15 minutes at 100°C. The d_{31} piezoelectric charge coefficients were derived from the planar coupling coefficients, k_p measured using an impedance analyser and the variation in d_{31} against poling field used to compare poling behaviour. Determination of the d_{33} piezoelectric charge coefficients for samples poled at the maximum field used ($4MVm^{-1}$) were made using a Berlincourt meter.

RESULTS AND DISCUSSION

A comparison of the densification behaviour for undoped and Mo^{6+} doped compositions sintered in air and oxygen are shown in Fig. 1 and table III.

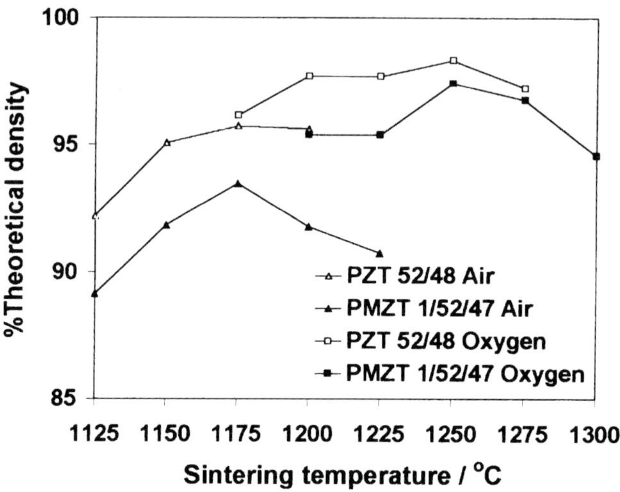

Figure 1: Comparison of densification behaviour of PZT 52/48 with PMZT 1/52/47 sintered in air and oxygen (assuming ρ_{theo} of $8Mg\ m^{-3}$).

It is apparent that increased densities were achieved from samples sintered in oxygen, compared with those sintered in air, which probably results from the removal of insoluble nitrogen which, when trapped within closed pores, inhibits densification during air sintering[12-14]. The temperatures necessary for optimum sintering in air are lower (typically by 50-100°C) than for materials prepared by conventional mixed oxide methods, also than those required to achieve maximum densification in oxygen in the present work. The lowering of densities observed for air sintered samples as the temperature exceeds that for optimum densification is probably caused by increasing pressure within closed pores (from the presence of nitrogen and lead vapour). In the case of oxygen sintered samples nitrogen would not be present and lead vapour pressure reduced[14], permitting continued densification. The densities for Mo^{6+} doped compositions are lower than for undoped material, as shown in table III, possibly as a consequence of the lead vacancies introduced for charge compensation.

XRD analysis, shown in Fig. 2, in particular the 110 and 111 peaks, indicates that the substitution of Mo^{6+} for Ti^{4+} shifts PZT towards the tetragonal side of the phase diagram[3], probably as a result of its small ionic size. All Mo^{6+} doped compositions show increased an temperature of the Curie point, see table III, compared with undoped compositions, especially those sintered in oxygen, which may indicate the reduction of Mo^{6+} in air sintered samples (possibly through a loss of oxygen). Alternatively increased lead loss (from higher sintering temperatures) and decreased oxygen loss, for samples sintered in oxygen, would limit the reduction of Mo^{6+} in order to maintain charge compensation.

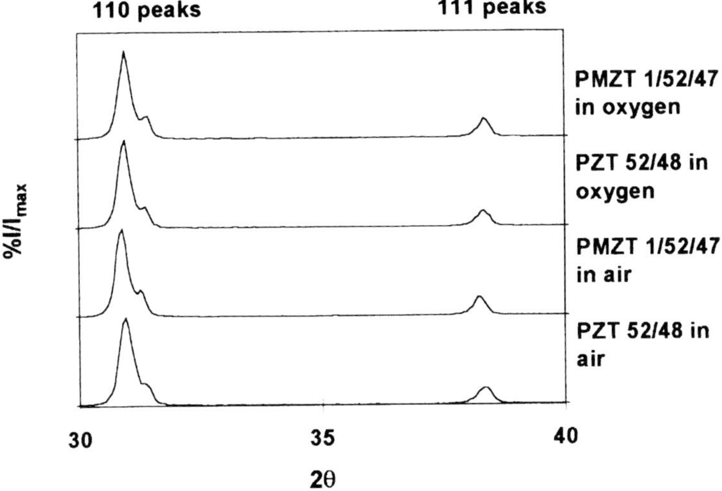

Figure 2: Comparison of XRD traces from PZT 52/48 and PMZT 1/52/47 sintered in air and oxygen.

Table III: Variation in density (assuming ρ_{theo} of 8Mg m^{-3}) and T_c for doped and undoped compositions prepared in this work.

Composition	Sinter T in air / °C	$\%\rho/\rho_{theo}$	T_c / °C	Sinter T in O_2 / °C	$\%\rho/\rho_{theo}$	T_c / °C
PZT 52/48	1175	95.73	392	1250	98.34	392
PMZT 1/52/47	1175	93.47	397	1250	97.44	408
PZT 57/43	1175	96.61	381	1250	98.42	379
PMZT 1/57/42	1175	92.42	385	1250	96.22	392
PZT 62/38	1175	96.27	368	1225	98.51	364
PMZT 1/62/37	1200	91.10	371	1250	96.74	376

The maximum poling field employed in this work was 4MV m^{-1}, above which problems from electrical breakdown were encountered. This field was sufficient to pole Mo^{6+} doped samples adequately, as shown in Fig. 3.

The poling studies, Fig. 4, indicate improvements in poling behaviour for Mo^{6+} doped specimens sintered in both air and oxygen, which can be explained in terms of increased A-site vacancies, which facilitate domain wall motion. The material may be described as having been 'softened'. This would also explain the enhanced final piezoelectric values, shown in table IV, for Mo^{6+} doped samples (PMZT 1/57/42 sintered in oxygen being the notable exception to this). Undoped samples sintered in an oxygen atmosphere also show improved poling characteristics, Fig. 4, compared with their air sintered counterparts and can be attributed to a reduction in oxygen vacancies, which 'harden' the material by pinning domain walls and increased A-site vacancies arising from the higher sintering temperatures. Piezoelectric properties were especially enhanced by oxygen sintering for PZT 57/43 compared with PZT 52/48, table IV, possibly indicating a shift in the tetragonal/rhombohedral phase ratio present towards that of the morphotropic phase boundary, which would alter poling behaviour.

The improved poling behaviour and higher piezoelectric coefficients (at 4MV m^{-1}) for the air sintered Mo^{6+} doped samples, compared with those for their oxygen sintered counterparts contrasts with that for undoped PZT and may be connected with the reduction of Mo^{6+} described earlier. It is, however, difficult to confirm whether this is the case during the present work.

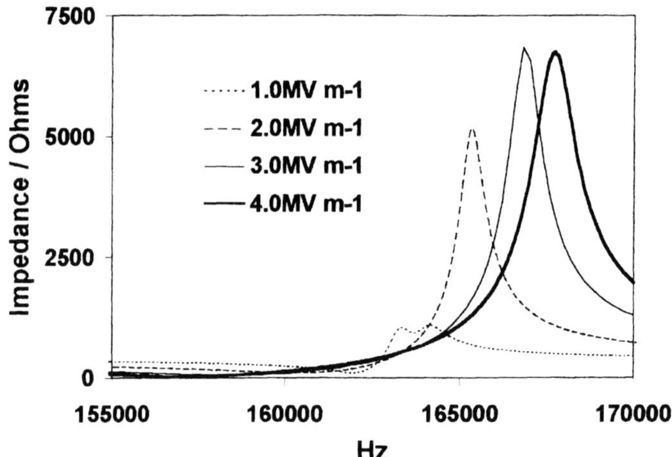

Figure 3: Change in impedance spectra with poling field, indicating poling is almost complete at 3MV m^{-1}, for PMZT 1/52/47 sintered in air.

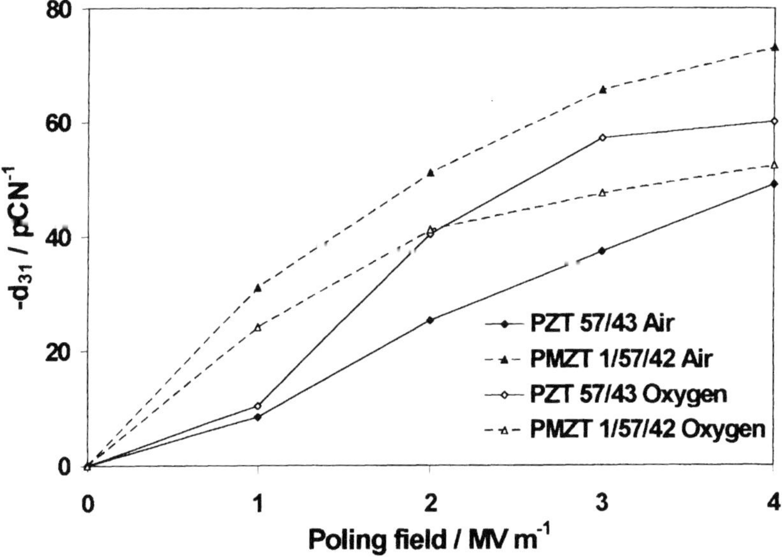

Figure 4: Variation of d_{31} with poling field for PZT 57/43 and PMZT 1/57/42 sintered in air and oxygen.

Table IV; Comparison of piezoelectric charge coefficients for samples poled at $4MVm^{-1}$ (all values in pCN^{-1}).

Composition	$-d_{31}$ air	$-d_{31}$ oxygen	d_{33} air	d_{33} oxygen
PZT 52/48	44	47	114	110
PMZT 1/52/47	62	52	138	138
PZT 57/43	49	60	83	125
PMZT 1/57/42	73	52	150	131
PZT 62/38	13	33	26	41
PMZT 1/62/37	47	40	108	108

CONCLUSIONS

Measurements of the piezoelectric properties of undoped and 1at% Mo^{6+} doped PZT, ceramics across a range of Zr/Ti ratios, prepared by a chemical route have been made. Increases in the Curie temperature associated with Mo^{6+} doping have been shown to be influenced by sintering atmosphere. Poling behaviour is observed to be improved by 1at% Mo^{6+} doping of PZT because of the presence of A-site vacancies which increase domain wall mobility. The 'softening' effects of oxygen sintering on undoped PZT, which can be associated with a reduction in oxygen vacancies, have been compared with the 'hardening' effect on Mo^{6+} doped PZT. Clearly further investigation is required before a mechanism for this can be elucidated.

ACKNOWLEDGEMENTS

The authors wish to record their thanks to Morgan Matroc Unilator for their assistance. They are also grateful to Mr T. P. Comyn for advice and aid with regards to dielectric and piezoelectric measurement. In addition Mr L. Blackwell would like thank EPSRC for their financial support.

REFERENCES

1. R.GERSON: 'Variation in Ferroelectric Characteristics of Lead Zirconate Titanate Ceramics Due to Minor Chemical Modifications', *Journal of Applied Physics*, 1960, **31** [1], 188-194.
2. N.P.PATEL, P.S.NICHOLSON: 'Comparison of Piezoelectric Properties of Hot-Pressed and Sintered PZT', *American Ceramic Society Bulletin*, 1986, **65** [5], 783-787.
3. B.JAFFE, W.R.COOK, H.JAFFE: *Piezoelectric Ceramics*, Academic Press Inc, London, 1971.
4. R.LAL, S.C.SHARMA, R.DAYAL: 'Piezoelectric Characteristics of Spray-Dried PZT Ceramics Modified by Isovalent, Supervalent and Subvalent Substitutions', *Ferroelctrics*, 1989, **100**, 43-55.
5. L.WU, T.S.WU, C.C.WEI, H.C.LIU: 'Dielectric Properties of Modified PZT Ceramics', *Journal of Physics, C: Solid State Physics*, 1983, **16**, 2803-2812.
6. L.WU, T.S.WU, C.C.WEI, H.C.LIU: 'Piezoelectric Properties of Modified PZT Ceramic', *Journal of Physics, C: Solid State Physics*, 1983, **16**, 2813-2821.
7. L.WU, T.S.WU, C.C.WEI, H.C.LIU: 'The D.C. Resistivity of Modified PZT Ceramics', *Journal of Physics, C: Solid State Physics*, 1983, **16**, 2823-2832.

8. D.L.SEGAL, 'Powders for Electronic Ceramics', *Electronic Ceramics,* B.C.H. Steele ed., Elsevier Science Publishers, Barking, England, 1991.
9. D.HIND, P.R.KNOTT: 'Aqueous and Sol-Gel Synthesis of Submicron PZT Materials and Development of Tape Casting Systems for Multilayer Actuator Fabrication', *British Ceramics Proceedings*, **52**, 1994, 107-119.
10. D.W.JOHNSON, Jnr: 'Non-conventional Powder Preparation Techniques', *American Ceramic Society Bulletin*, 1981, **60** [2], 221-224, 243.
11. D.W.JOHNSON, Jnr: 'Innovations in Ceramic Powder Preparation', *Advances in Ceramics,* **21**, *Ceramic Powder Science*, The American Ceramic Society Inc, 1987.
12. G.S.SNOW: 'Fabrication of Transparent Electrooptic PLZT Ceramics by Atmosphere Sintering', *Journal of the American Ceramic Society*, 1973, **56** [2], 91-96.
13. G.S.SNOW: 'Improvements in Atmosphere Sintering of Transparent PLZT Ceramics', *Journal of the American Ceramic Society*, 1973, **56** [2], 479-480.
14. J.H.MOON, H.M.JANG: 'Effects of Sintering Atmosphere on Densification Behaviour and Piezoelectric Properties of $Pb(Ni_{1/3}Nb_{2/3})O_3$ - $PbTiO_3$ - $PbZrO_3$ Ceramics', *Journal of the American Ceramic Society*, 1993, **76** [2], 549-552.

ADDITIVE-FREE FERROELECTRIC THICK FILMS

DH PEARCE, TW BUTTON

Functional Materials Group, IRC in Materials, University of Birmingham, Edgbaston, Birmingham B15 2TT, UK.

Email d.h.pearce@bham.ac.uk

ABSTRACT

Much of the limited previous work carried out in the area of thick film processing of piezo- and pyroelectric films, and other ferroelectric materials such as strontium barium titanate, has been in line with conventional thick film technology, i.e. with the main aim being to obtain densification at the normal processing temperatures of around 800 to 900 °C. This has only been possible in most cases through the utilisation of either a glass frit to allow interparticular bonding, by additions of excess PbO as a liquid phase sintering agent, or more recently through the use of a eutectic phase of PbF_2/PbO. All these routes enable densification to be achieved to some extent, and at temperatures much lower than those commonly used for bulk sintering of PZT, $Pb(Zr_xTi_{1-x})O_3$ (typically 1100 -1300°C). However, the electrical properties of PZT are known to be strongly dependant on the microstructure of the material, and in particular on any non-ferroelectric grain boundary phases present. For example, excess lead oxide at the grain boundaries is deleterious to the piezoelectric properties, as are other glass phases.

In this paper alternative methods of producing good quality PZT thick films are described which do not utilise microstructural binding additions. The processing routes involved in fabricating phase pure PZT thick films are described. Dense films have been obtained on platinised alumina substrates, sintered at temperatures between 1100 and 1200 °C. The effect of the bottom electrode material on the quality of the film, and the effect of processing on the substrate are outlined and explained. In processing of these films, a doped zirconium titanate phase is first deposited and partially densified at high temperature, followed by the addition of lead oxide at a lower temperature in order to form PZT. The many variables involved in this route, and the potential for optimisation, are described. The effects of the different processing routes on the microstructure and electrical properties of the sintered films are reported, and potential for future work outlined.

INTRODUCTION

PZT has for many years been utilised in its bulk form as a material for a wide range of transducers, utilising its large piezoelectric effect. The vast majority of PZT is still manufactured by conventional technical ceramic methods, resulting in, for example, simple puck shapes for applications such as ultrasonic transmitters or gas igniters. Other techniques though have uses in more precise or small scale applications, examples being in non-volatile memories and multi layer actuators, for which thin film and thick film (tape casting) technologies are utilised. New types of devices have arisen from these techniques, in particular thin film technology, which has received a great deal of attention over the past decade. The more simple technique of thick film processing, in the form of screen printing has, however, received much less attention. This is perhaps mainly due to the limited number of applications where printed films of PZT are currently envisaged to be utilised, particularly when applied on rigid alumina substrates. Applications suggested for printed PZT films have covered miniature accelerometers/ mass sensors, micro actuators, SAW devices and pyroelectric sensors[1,2], but devices based on screen printed PZT thick films have as yet not been utilised in commercial devices.

Conventional thick film technology is based on a standardised processing route, to enable a wide range of functional films to be laid down on standard ceramic substrates, which are commonly 96% alumina. The route follows that illustrated in Figure 1. Many different types of materials can be processed via this route, due to its flexibility through the choice of polymeric and vitreous binding additions. The need to maintain the processing temperature, however, means that often large additions of the vitreous binder are necessary in order to obtain a dense film. For some thick film applications, most notably superconductors, additions of this kind are not acceptable, since any post processing grain boundary phase will deteriorate the properties and most likely prevent superconductivity in the film. Indeed, such films are not in most cases truly sintered, and are merely bound together with the vitreous additive, with the maximum extent of sintering being the initial stage of particle contact. In addition to this, the contact between the functional particles will reduce the driving force for the network of particles within the liquid phase addition to densify.

Previous work has shown that novel routes can be discovered which avoid the need for glass frit additives in functional ceramic thick films, resulting in extremely good properties for the sintered film[3,4]. This type of approach was therefore undertaken in this work for the fabrication of other functional ceramic thick films. The processing route undertaken for this work was derived from previous work in the area of reactive calcination for powder technology[5].

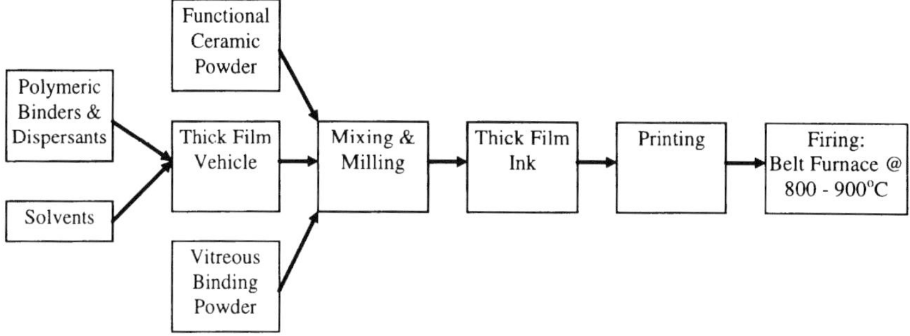

Figure 1 - Typical Thick Film Formulation, Printing and Firing Schedule

EXPERIMENTAL PROCEDURE

Ink Preparation

Two ink formulations were made, from stock ceramic powders. The starting materials used were MEL E101 zirconia, Tioxide HPT1 titania, Fluka lead oxide and Fluka lanthanum oxide powders. The zirconia, titania and lanthanum oxide powders were weighed out in the proportions which were consistent with a 2% doped PZT with a 52:48 zirconia:titania ratio, the point at which the piezoelectric properties peak in the $PbZrO_3$ - $PbTiO_3$ phase diagram[6]. This mixture was mixed with a standard Terpineol-based vehicle and milled on a three roll milling machine until deagglomerated. A similar ink was made consisting of pure lead oxide in the same vehicle.

Electrode inks were formulated, based on platinum and silver. A commercial pure platinum ink (Gwent Electronic Materials 72% Pt) was used for the bottom electrodes, in some cases modified with additions of E101 zirconia to improve adhesion. A silver flake based ink was formulated for the top electrodes, consisting of Johnson Matthey FS2 silver flake and a 5% addition of a powdered lead borosilicate glass frit bonding agent, which had been previously milled to reduce its particle size to less than 5 microns. This was also mixed with the standard thick film vehicle and subjected to three roll milling.

PRINTING AND SINTERING

A set of three screens were designed and fabricated for the creation of a thick film test structure, to enable microstructural and electrical analysis. This is shown schematically in Figure 2, which shows an exploded arrangement of these patterns. The platinum electrodes were printed, dried at 40-80°C and sintered at 1200-1400°C, arranged upright in a muffle furnace to prevent warping of the 96% alumina substrate. On top of this was printed a layer of the zirconium titanate (ZT) film, typically 25 microns thick. This was sintered in the same way, again at a temperature between 1200 and 1400°C. On top of the sintered ZT print a film of the lead oxide ink was then applied. This was fired at 900°C in an enclosed crucible, to reduce furnace contamination and lead oxide volatilisation, and to allow the PbO to diffuse into and react with the ZT layer to form the correct phase of PZT.

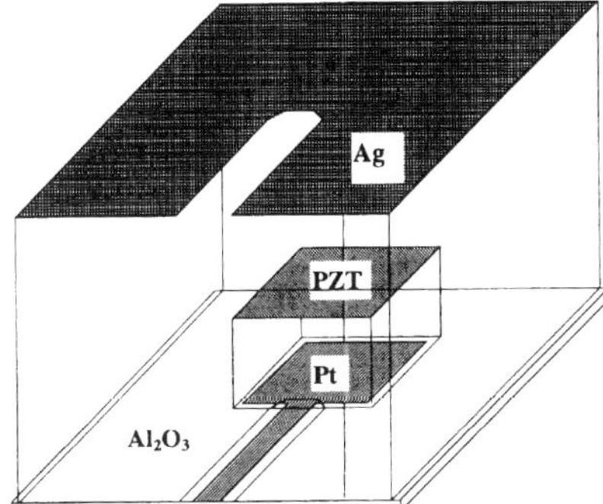

Figure 2 - Screen patterns for printed PZT test structure.

RESULTS

The success or otherwise of the processing route was analysed initially through a combination of microstructural and XRD analysis. For the microstructural analysis, a uniform dense film at the final stage of processing was to be the judge of a successful thick film. X-Ray measurements were taken at each stage of processing to determine at which point the desired phases appeared, and the observation of a single phase PZT film was the desired aim. Finally, electrical analysis would be used to concentrate on the actual electrical behaviour of the materials produced, which would be a more sensitive measurement than the previous two.

XRD Analysis

The progression of the separate zirconia and titania phases in the initial printed film to that of a single phase corresponding to zirconium titanate is shown in Figure 3. The three sintering temperatures shown were held for one hour in each case. At 1200°C there is no apparent reaction between the components, only a certain degree of peak shifting, indicating a small amount of interdiffusion, or possibly a reaction with the silica or other impurity in the substrate material. At 1300°C, however, the reaction appears to have proceeded to completion, with a sharply defined main peak at $2\theta=35.5°$. With a further increase to 1400°C this peak shifts to 30.4°, likely to be a result of substrate interactions, together with evidence of a new phase by the appearance of a new peak at 28.4°.

Reacting the ZT prints further with printed PbO films resulted in the diffraction patterns shown in Figure 4, presented together with a standard trace for a partially sintered thick film of pre-reacted PZT (PZT-5A, Morgan Matroc TPD, Southampton) for comparison. Clearly, PZT is being formed from all three pre-reacted films, but the broad peaks in all cases compared to the sharper peaks for the pre-reacted PZT indicate compositional inhomogeneities, probably due to the uncontrolled quantities of lead oxide added to the ZT film.

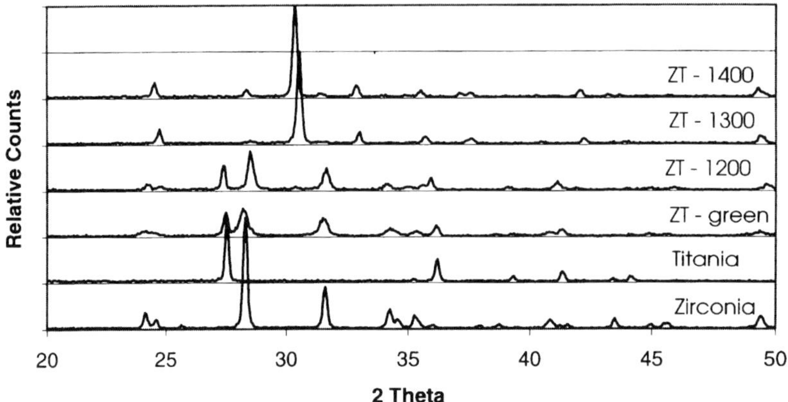

Figure 3 - Development of the ZT phase print throughout processing.

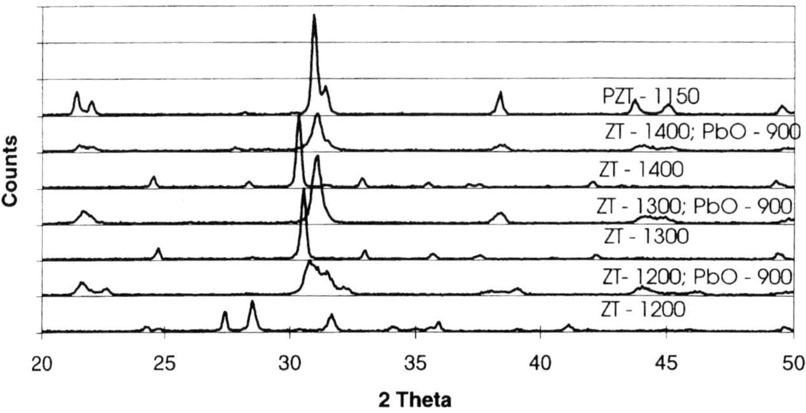

Figure 4 - Reaction of lead oxide at 900°C with previously reacted ZT prints.

Microstructural Observations

In order to elucidate the measurements carried out through XRD, SEM analysis was performed on fracture surfaces of samples at stages throughout the processing route. Illustrating the effect of the varying processing temperatures performed on the ZT film, Figure 5 and Figure 6 show the appearance of films which had been pre-reacted at 1200 and 1300°C respectively. After pre-reaction at 1200°C a more open and uneven film structure is obtained than reaction at 1300°C. This is explained by the simultaneous reaction and sintering of the ZT mixture not taking place until the higher temperature, but allowing sufficient remaining volume for the PbO intrusion and diffusion to form the final PZT phase. A close match therefore between the ZT reaction and the PbO intrusion has been achieved for the

1300°C reacted film. As evidenced by the XRD results, however, this film does show clearly the incomplete reaction and/or phase inhomogeneity, illustrated in Figure 7. This shows the surface of the film, and the appearance of a glassy-like phase separating the PZT grains, which will be the remaining lead oxide in the film.

Electrical Characterisation

After sintering, films were printed with the formulated silver ink and sintered at 600°C to form a bonded top electrode to solder electrical connections on to. For both the 1200 and 1400°C ZT films poling of the films was not carried out. In the case of the 1200 film, the open structure allowed many conductive paths through the material to the bottom electrode, and for the 1400 film the sintering shrinkage of the ZT film generated defects in the film surface, resulting in shorting. For the 1300 film, however, poling could be carried out, and was performed under normal conditions as for bulk PZT, i.e. with an applied field of 2.5 kV/mm in an oil bath at 130°C for 10 minutes. The capacitance of the film was measured at 1kHz with a 1V signal, and from this the relative dielectric constant ε_r determined. For this film, this was determined to be around 200, compared to a typical value for a bulk soft doped PZT material of 1500. Also, no resonance peaks were visible over the entire range of the analyser (10^2 to 10^6 Hz), though this is most likely due to the film volume being very small compared with the substrate, resulting in a very large flexural and lateral constraint on the film. Piezoelectricity was, however, evidenced by the fact that the device produced audible sound during testing.

CONCLUSIONS

A novel technique for applying PZT films on to standard alumina substrates has been developed, where the lead oxide phase is introduced at the final stage and therefore at a lower temperature than normal for bulk PZT processing. XRD results indicate that the correct PZT phase is produced when lead oxide is reacted at 900°C with a zirconium titanate film which has been pre-reacted at 1300°C. This material has been shown to exhibit piezoelectricity, though with a much reduced dielectric constant compared with normal phase pure PZT. Further refinements to the process, involving closer attention being paid to the separate reaction stages of the process, together with control of the amount of lead oxide printed on the ZT film, should allow for improvements on the results obtained thus far. In future this method should be a viable route for the deposition of pure, non-fritted, thick films of PZT, and eventually will show properties similar to those of bulk materials.

ACKNOWLEDGEMENTS

This work was carried out as part of an EPSRC ROPA award on thick film processing of functional ceramics. The authors also acknowledge the assistance Professors Loretto and Harris of the IRC in Materials and the School of Metallurgy and Materials respectively, for the use of departmental facilities.

REFERENCES

[1] N.M. WHITE and J.D. TURNER: "Thick-film sensors: Past, present and future", *Meas. Sci. Technol.*, **8** (1), 1-20, 1997.

[2] C. LUCAT, F. MENIL and R. Von Der MUHLL: "Thick-film densification for pyroelectric sensors", *Meas. Sci. Technol.*, **8** (1), 38-41, 1997.

[3] T.W. BUTTON, N. McN ALFORD, F. WELLHOFER, T.C. SHIELDS, J.S. ABELL and M. DAY, *IEEE Mag.*, 1991, **27**, 1434-1440, 1991.

[4] T.W. BUTTON, P.A. SMITH, Z. WU, N. McN. ALFORD, L.E. DAVIS and S. PENN, *IEEE Trans. App. Superconductivity*, 1993, **3**, 1450-1455, 1993

[5] S. KIM, G.S. LEE, T.R. SHROUT and S. VENKATARAMI: "Fabrication of fine-grain piezoelectric ceramics using reactive calcination", *J. Mat. Sci.*, **26**, 4411-4415, 1991.
[6] D. BERLINCOURT: "Piezoelectric ceramic compositional development", *J. Acoust. Soc. Am.*, **91** (5), 3034-3040, 1992.

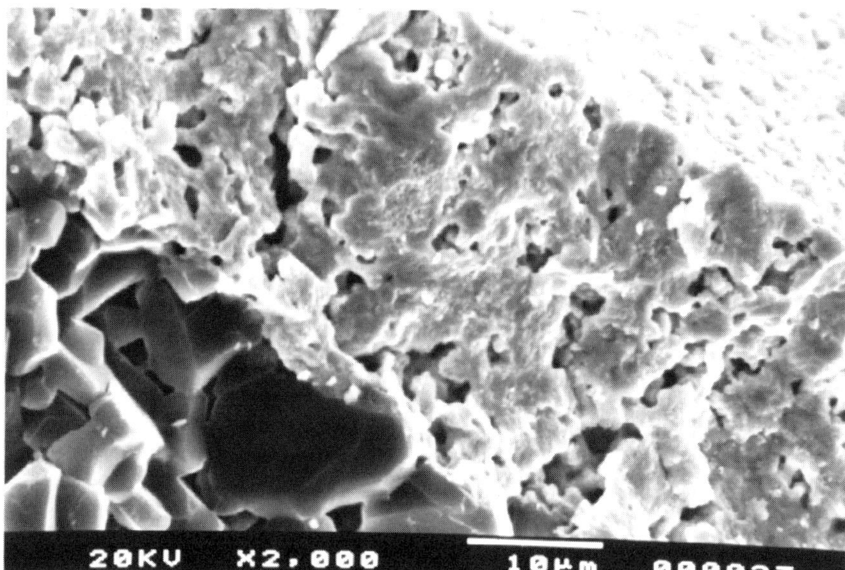

Figure 5 - SEM micrograph of PbO reacted ZT film, pre-reacted at 1200°C

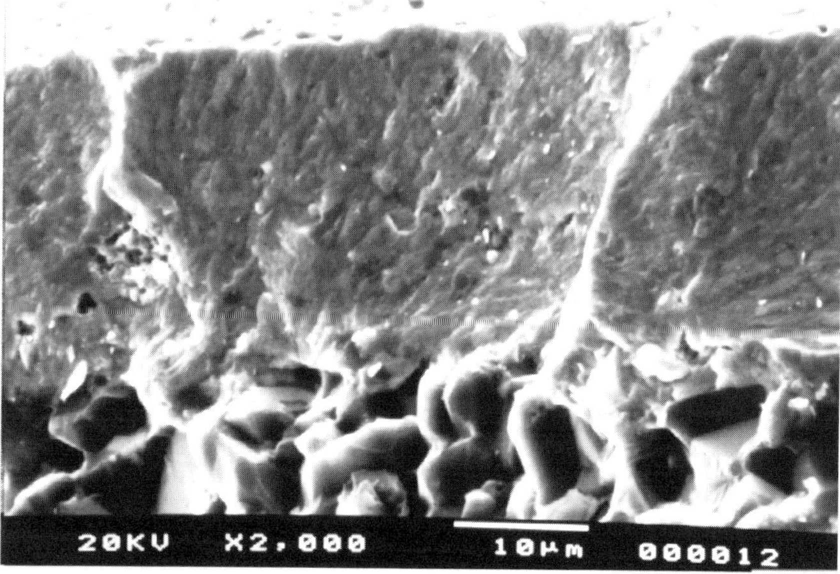

Figure 6 - SEM micrograph of PbO reacted ZT film, pre-reacted at 1300°C

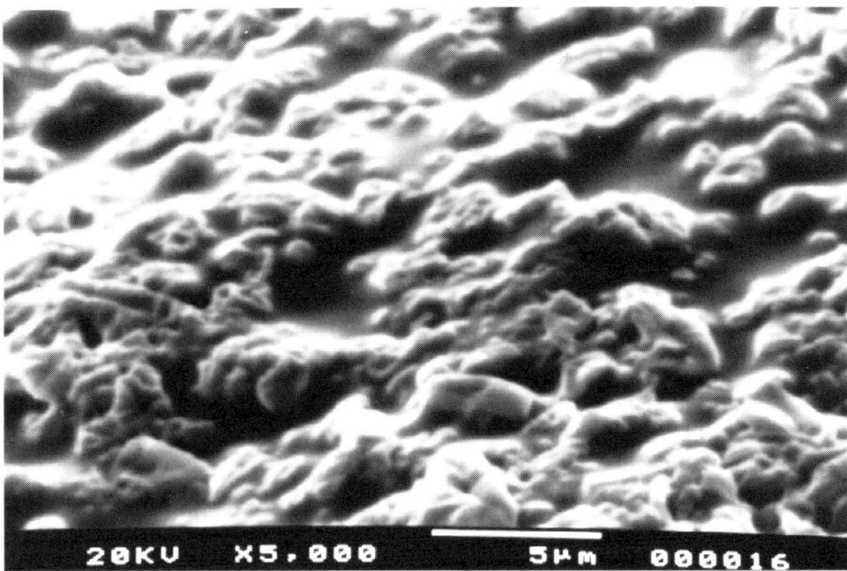

Figure 7 - SEM micrograph of surface of PbO reacted 1300 ZT film.

THE BEHAVIOUR OF PIEZOELECTRIC TUBULAR STRUCTURES FOR NOVEL DEVICES

J.E.HOLMES, D.H.PEARCE, AND T.W.BUTTON
IRC in Materials and School of Metallurgy and Materials, University of Birmingham
Email J.E.Holmes@bham.ac.uk

ABSTRACT

Advanced ceramic processing techniques have been used to fabricate complex tubular piezoelectric structures with potential application in a wide range of new sensor and actuator devices. The operation of these structures is currently incompletely defined and an analysis tool is required to characterise these systems more fully. For the realisation and optimisation of novel functioning PZT devices the origins of the electromechanical responses need to be understood. The development of a finite element model as a flexible analysis tool for applying known material properties and parameters to complex geometries is described. Tubular lead zirconate titanate ceramic structures in the form of helical springs have been fabricated and characterised in terms of their voltage-extension behaviour. The experimental observations are compared to predictions from the finite element model. Possible origins of the observed effects are proposed, and routes to the realisation of new devices are discussed.

INTRODUCTION

Piezoelectric phenomena have attracted much attention in recent years as the applications grow in various fields such as transducers and filters[1]. This paper is primarily concerned with the characterisation and analysis of a helical tubular piezoelectric structure, and although the possibility of forming a piezoelectric material into a spring has been considered before it has hitherto not been realised[2]. Ceramic processing technologies currently being developed at the University of Birmingham enable the fabrication of net shape components with complex geometries and enhanced mechanical properties[3]. Helical structures promise the possibility of creating piezoelectric transducers which operate in the millimetre deflection range and hence widen the applications to which piezoelectric ceramics may be applied. A spring also operates essentially as a spring mass system offering the possibility of sensors with low operating frequencies such as used in seismic sensing elements.

The concept of the helical device was as a flexible sensor, the tubular section operating as a straight tube, and the helical structure merely adding flexibility to the device. When the device was formed it was found to produce a voltage upon compression and over large deflections. It was also found to produce the same sign of voltage output whether under compression or tension.

The study of vibrations within simple piezoelectric structures of different geometries has been an important necessity for the development of devices and applications. Previous authors have investigated the operation of piezoelectric cylinders of both infinite and finite length, both in axisymmetric and flexural free vibration[4]. However, due to the convoluted shape involved in the present study the methods of solution previously employed are inappropriate and, therefore, finite element analysis techniques have been employed. This paper describes the development of an adaptable linear model to understand the behaviour of the springs and to aid the eventual development of sensor and actuator devices. The predictions from the model are compared to the results of a preliminary electromechanical characterisation of real structures.

FABRICATION ROUTE

The device requires a ceramic tube to be sintered in the form of a helical spring, with electrodes on both the inner and outer surfaces of the tube. It is not possible to machine such a device, so a net shape fabrication method is required. The final device must be strong enough to withstand the high uniform stresses that it will experience as a spring, and although the electrodes do not have to possess an extremely low impedance, good coverage over the entire internal and external surfaces is desired.

The springs were made using a Viscous Processing (VP) route, the methods for which are detailed elsewhere[5]. In brief, lead zirconate titanate (PZT) powder (PZT5A) was mixed with polymers and water on a twin roll mill to form a viscous dough. A simple ram extruder was used to extrude the dough through a tube die which was then coiled into a helical spring. After drying, burnout of the polymer was carried out at 1 °/min up to 600°C, followed by sintering at 1200°C for 1 hour in an enclosed crucible. Typical dimensions of the sintered tubes were in the range OD = 1-3mm and ID = 0.5-2mm. Overall dimensions of the springs were in the range OD = 10-50mm and lengths up to 200mm. Electrodes were applied to the internal and external surfaces of the tubular structure using a specially developed thick film fritted silver ink followed by sintering at 600°C. Poling was carried out in a mineral oil bath at 130°C for 10 minutes by applying a field of 2.5kVmm^{-1} across the tube wall.

FINITE ELEMENT ANALYSIS

Introduction

Finite element analysis has been used for the initial investigation into the operation of the spring due to the flexibility it allows for geometric shape, the information it supplies regarding the stress state within the body of the structure and the availability as commercial software packages. All analysis performed to date has used ABAQUS. The constitutive relations used by ABAQUS in the analysis are;

$$D = \varepsilon^S E + eS \tag{1}$$

$$T = -e_t E + c^E S \tag{2}$$

where the independent variables are the electric field (E) and strain (S), the dependant variables are electric displacement (D) and stress (T), the constants used are permittivity (ε), the piezoelectric stress constant (e) and the elastic stiffness constant (c). All constants were assumed to be those for standard PZT 5A material [6].

ε

The Behaviour of Helical Springs

Many textbooks on elasticity discuss the simple operation of a tension spring, Figure 1 shows a simple diagram of a spring. The prominent effect in a close-coiled helical spring is torque (T) and as the angle α increases a bending moment (M) becomes apparent, the relationship between these effects and the angle α is given in equations (3) and (4).

$$M = FR\sin(\alpha) \tag{3}$$

$$T = FR\cos(\alpha) \tag{4}$$

A more complete description of the operation of a spring may be found in Reference [7]. For a poled polycrystalline ceramic (symmetry class 6mm) shear forces do not produce a change in polarisation in the poled direction, a bending moment applied to tube produces principle stresses but their effects cancel out around the tube.

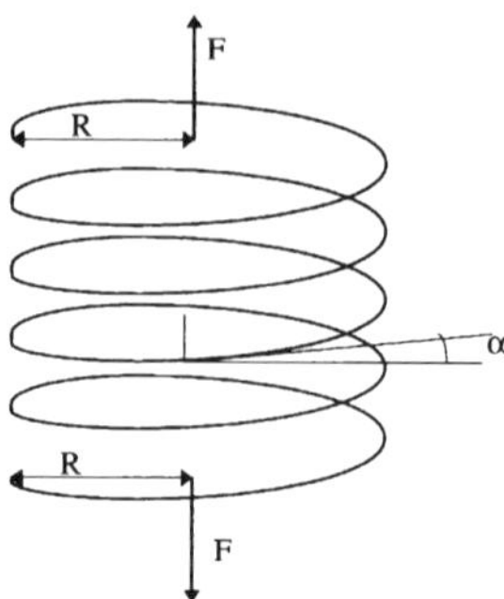

Figure 1 - Diagram of a close coiled helical spring, showing the major radius R, the applied force F and the pitch angle α.

The Finite Element Model

The spring was produced from a tube poled radially, so when producing the finite element model a local co-ordinate system was defined at each element to represent the material orientation. The mesh and orientated material parameters are produced from a separate program before being submitted to ABAQUS for analysis. The model was generated using 20-node quadratic elements. A typical mesh for a quarter turn of a spring can be seen in the output from ABAQUS in Figure 2. Also shown in the figure is the element numbering nomenclature that is used in the subsequent discussion. The position going around the tube is numbered clockwise from the point nearest the axis of the spring, and the through-thickness position is numbered from the inner to outer surfaces. In producing the mesh the number of elements used was determined purely by trial and error until the predicted output voltages and displacements converged. A zero potential boundary condition is placed upon the nodes on the interior of the tube, the nodal points on the exterior have equipotential, thus simulating the electroded surfaces.

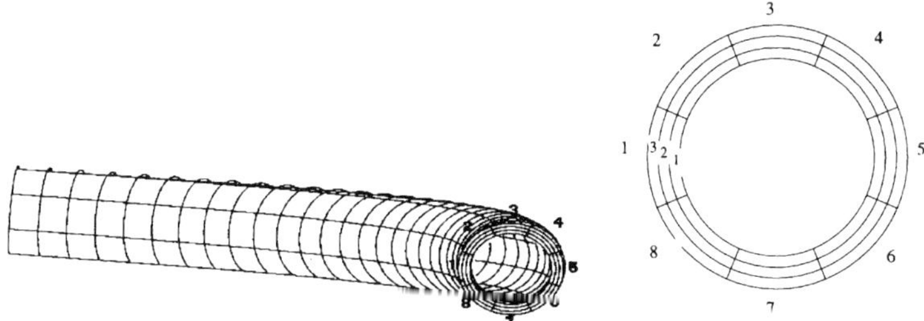

Figure 2 - Mesh generated to model a ceramic spring and the nomenclature used for results.

In approaching this study much attention had to be placed on the method of actuating the spring. The simplest method is to examine the fundamental vibration and study the stress states within the material, and this has been done for the spring alone and for a spring-mass system, the mass being represented by a point mass with suitable rotary inertia. Rigid elements connected the end of the spring to the centre of mass, which also enabled a set torque to be applied to the axis of the spring as shown in Figure 1.

The torsional component of the stresses throughout the cross-section may be seen in Figure 3. The stress shown is for a two turn spring with an overall diameter = 16mm and comprising of a piezoelectric tube with OD = 1.6 mm, and ID = 1.2 mm at a pitch angle of 0.04 radians, with a force of 0.5 N applied along its axis. The torsional shear component can be seen to depend upon the distance from the centre of the tube as expected, with the shear stress increasing from the inside to the outside of the tube. The variation of the torsional component around the tube comes from the combined effect of the torque applied to the spring and transverse shear.

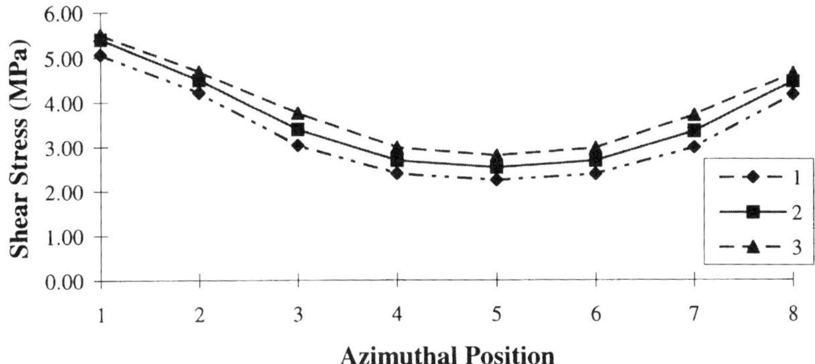

Figure 3 - Variation in torsional shear stress around the cross-section of the tube.

The variation in stress along the axis of the tube is shown in Figure 4. It can be seen that the stress on the inner surface of the spring near its axis (azimuthal positions 2 and 8) is greater than that near the outer surface (azimuthal positions 4 and 6), due to the greater curvature of the tube. The model predicts forces which would suggest a net stress through the axis of the tube, but the resultant output voltage would be dependent upon the direction of the applied force, in contrast to the observations of actual devices.

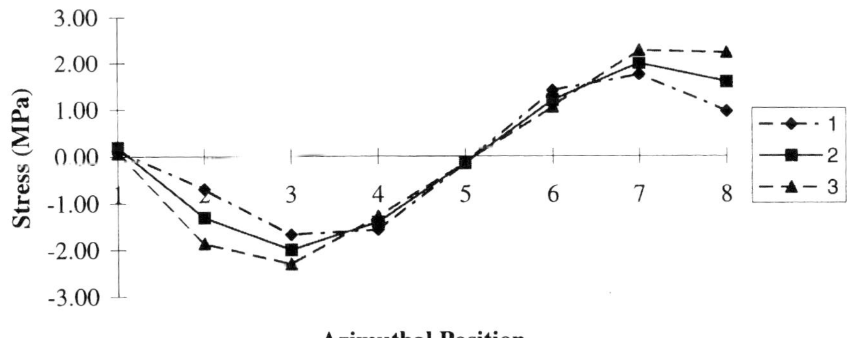

Figure 4 - Stress in the direction of the tube.

THE BEHAVIOUR OF REAL DEVICES

A preliminary characterisation of the electromechanical properties of a piezoelectric tubular spring has been carried out, comprising measurements to test the linearity of the voltage-extension response. A spring with an overall diameter of 16mm and comprising 15 turns of piezoelectric tube with OD = 1.6 mm, and ID = 1.2 mm at a pitch angle of 0.04 radians was used for these experiments. In order to apply loads along the axis of the spring low-mass plastic end caps were glued to each end. A simple experiment was set up using a load frame to stress the spring to a pre-set compression, followed by cycling through a compression of 0.2 mm peak to peak at a frequency 3Hz, the experimental arrangement is shown in Figure 5.

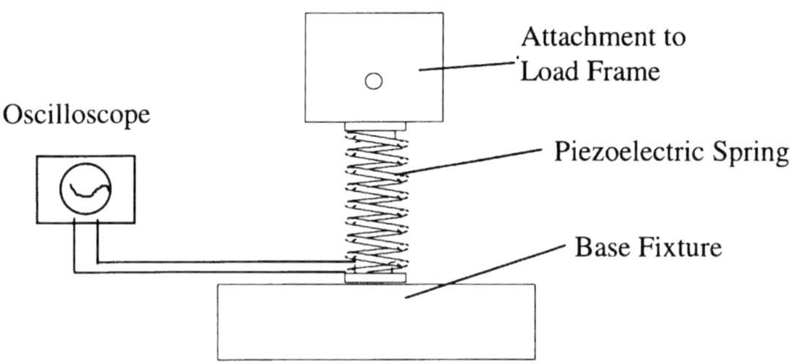

Figure 5 - Schematic showing the experimental set-up for the initial investigation into the linearity of the response of the spring.

The results are shown in Figure 6. The zero compression point is taken as the natural length of the spring under the combined effect of its own self-mass and that of the end caps. The measured peak-to-peak voltage ranged from 90-160mV for the range of pre-compressions investigated, in contrast to a value of 2.4 mV predicted by the model for this configuration. It is also evident that the measured output voltage varies in a non-linear manner with respect to the pre-compression of the spring.

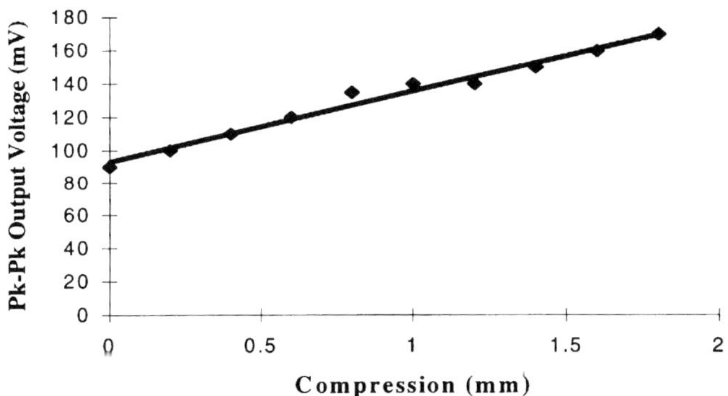

Figure 6 - Variation in output voltage with applied pre-compression.

DISCUSSION

At the beginning of this work a number of possible origins of the behaviour of the spring device were considered. For example, piezoelectric materials are anisotropic, and the anisotropy is introduced by poling. For the case of the spring this has been assumed to be radial i.e. in the direction of poling. There may, however, be some anisotropy in the sintered ceramic before poling, which may affect the final sample properties. Another possibility is that the high strain associated with the operation of the spring could change the polarisation. PZT is ferroelectric i.e. an electric field can align the dipole moments of the ceramic. However under high stresses, like a ferroelastic material, the dipole orientation may change and the change in anisotropy could account for the observed behaviour. However, this effect would be expected to increase with extended operation and this has not been observed.

During bending or torsion the strains induced will cause deformation of the cross-section which may lead to an average strain through the cross-section. Of particular note is the ovalisation which is commonly observed in the plastic deformation of tubes. If such an ovalisation is significant then a strain in the azimuthal direction of the cross-section of the spring will be observed. This effect would be independent of compression or elongation. However, finite element modelling conducted on straight tubular sections subjected to bending moments, and allowing for non-linear geometrical changes, does not suggest a significant effect due to a cross-sectional change.

Another possible cause is that the strain induced in the material of the spring may be large enough for the higher order piezoelectric terms to become significant. These are not normally considered in conventional devices but the output would be independent of the sign of the stress - compression or elongation. At present this would seem to be the most likely explanation of the observed responses as a ceramic body with a spring geometry is subjected to large stresses.

The measurements conducted so far indicate that the response from the device is highly non-linear. For sensor applications this effect could be utilised by pre-loading the spring to produce an increased output voltage. An interesting observation of the current structure is that applying an alternating voltage across the electrodes produces axial movements of the spring in the millimetre range. Of course, the current electrode configuration provides little actuation force, but consideration should be given into utilising this stress directly by manipulation of the position and geometry of the electrodes. One possibility would be to split either the inner or outer electrode surfaces in order to allow the potential difference between the upper and lower sections to be utilised as shown by the stress variation in Figure 4. In order to utilise the predominant, torsional, stress a more elaborate electrode configuration is desired, either on the surface[2] or more effectively within the body of the ceramic, perhaps also using non-circular helical sections. Work is currently in progress to attain these goals.

CONCLUSIONS

Novel helical piezoelectric structures comprising sintered PZT tubes have been fabricated and characterised. Finite element models have been developed to investigate the origin of the electromechanical response, and these currently predict much smaller output voltages than those observed experimentally. Preliminary electromechanical characterisation indicates a non-linear response with extension. For applications as a sensor this effect may be utilised by operating the spring in a pre-strained regime. The present ceramic structure and electrode configurations do not lend themselves to actuator applications, but work is currently in progress to realise the electrode configurations that would produce viable actuators and more efficient sensors [8,9]. The present model can be easily adapted to consider the possible configurations and will be used in subsequent analysis and optimisation of these new devices.

ACKNOWLEDGEMENTS

The author would like to thank EPSRC and Basys Marine Ltd. for supporting this work. Thanks are also due for the technical assistance of G. Dolman and C. Meggs of the Functional Materials Group throughout this work.

REFERENCES

1. T.IKEDA: *Fundamentals of Piezoelectricity*, Oxford University Press, Oxford, 1990.
2 R.ADLER: Torsional Ceramic Transducer, United States Patent, 1975, Patent No. 3,900,748
3. D.H.PEARCE, G.DOLMAN, P.A.SMITH and T.W.BUTTON: 'Colloidal Processing of Lead Zirconate Titanate (PZT)', *Electroceramics V*, J L Baptista, J A Labrincha and P M Vilarinho ed., Proceedings of the International Conference on Electronic Ceramics and Applications, 2-4 Sept 1996, Aveiro, Portugal, Book 2, 385-388.
4. H.S.PAUL and K.NATARAGAN: 'Flexural Vibrations in a Finite Piezoelectric hollow cylinder of class 6mm', J. Acoust. Soc. Am., 1996, **99** (1), 373-382.
5. N.ALFORD, J.D.BIRCHALL and K.KENDAL: 'High-strength ceramics through colloidal control to remove defects', Nature, 1987, **330** (6143), 51-53.
6. R.HOLLAND: 'Accurate Measurements of Coefficients in a Ferroelectric Ceramic', IEEE Transactions on Sonics and Ultrasonics, 1969, **16** (4), 173-181.
7. C.J.ANCKER and J.N.GOODIER: 'Pitch and Curvature Corrections for Helical Springs', Transactions of the ASME - Journal of Applied Mechanics, 1958, **25**, 466-495.
8. D.H.PEARCE and T.W.BUTTON: 'Processing and Properties of Ag/PZT Composites', to be presented at the 29th Conference and 4th UK Transducer Materials and Transducer Workshop 'Anisotropic Dielectrics and Ferroelectrics', 6 – 9 April 1998, Canterbury, and published in Ferroelectrics.
9. D.H.PEARCE and T.W.BUTTON: 'Multi-Phase Piezoelectric Structures through Co-Extrusion', to be presented at the Institute of Materials Ceramic Convention, Cirencester, 21-22 April 1998, and published in Brit Ceram Trans.

STRUCTURE-PROPERTY RELATIONSHIPS IN STRONTIUM BARIUM NIOBATE CERAMICS.

P. M. Arrondelle *, D. Hind *, N.W. Thomas** and R.M. Brydson*

*Department of Materials, School of Processing Environmental & Materials Engineering, University of Leeds, Leeds LS2 9JT. Email: metpma, cerdah and mtlrmdb@leeds.ac.uk
** Watts Blake Bearne & Co. Limited, Park House, Courtenay Park, Newton Abbot, Devon TQ12 4PS. Email: nthomas@wbb.co.uk

ABSTRACT

Strontium Barium Niobate Ceramics, $Sr_xBa_{1-x}Nb_2O_6$, represent an important class of ferroelectric ceramics. The focus of this study is on Ba-rich samples with x = 0.30 and 0.35, and varying degrees of (Ti, W) co-doping, thus the composition the formula becomes $Sr_xBa_{1-x}Nb_{2-2y}Ti_yW_yO_6$, with $0.00 \le y \le 0.200$. Materials were prepared by a conventional mixed oxide route, but with the period of calcination increased to 24 hours in order to encourage thermodynamic equilibrium between the precursors.

Novel results are presented on the stability of the parent tetragonal tungsten bronze phase (TTB). Extended calcination times appear to allow other phases to form which have been characterised by SEM/EDX, EPMA, TEM, XPS and XRD.

The relationships between crystal structure, microstructure and dielectric properties (ε_r vs. T) are highlighted, with an assessment given of future research directions in this system.

INTRODUCTION

Strontium barium niobate (SBN), $Sr_xBa_{1-x}Nb_2O_6$, has been intensively investigated for its ferroelectric properties for values of x between 0.25 and 0.75. This compositional range is usually referred to as SBN100x, i.e. SBN25 to SBN75. For these values it exhibits the tetragonal tungsten bronze (TTB) structure, space group P4bm. The dielectric response of SBN is a function not only of temperature but also of the frequency of the applied measuring field, and hence it belongs to the relaxor class of ferroelectrics. The dielectric properties are also a function of the Sr/Ba ratio: as it rises, the temperature of maximum relative permittivity tends to fall; the relative permittivity increases as does the degree of frequency dispersion.

Both Francombe [1] and Ismailzade [2] are credited as being the first investigators of the material in 1960; Carruthers and Grasso investigated the $SrO-BaO-Nb_2O_5$ phase diagram[3], and the majority of the prior research carried out concentrated on the high Sr content compositions as these have been the ones found to have the most promising properties [4-11]. The present work details studies of high Ba content compositions, and the effects of both co-substituting titanium (Ti) and tungsten (W) on to the niobium (Nb) sites and modifications to the methods of fabrication.

EXPERIMENTAL

High purity grades of $BaCO_3$ (99.997%), $SrCO_3$ (99.994%), Nb_2O_5 (99.999%), TiO_2 (99.995%) and WO_3 (99.998%) sourced from Alfa/JM were micronised in propan-2-ol for 30 minutes and calcined at 1200°C for 24 hours. The resulting powder was characterised by XRD and TEM; pellets were pressed in a 15mm die at approximately 70MPa and sintered at 40°C below their experimentally determined melting point (for example 1270-1340°C for x = 0.30 compositions) for 4 hours. The values of x examined were 0.30 and 0.35 (SBN30

and SBN35), values of y were 0.000, 0.005, 0.010, 0.015, 0.020, 0.050, 0.075, 0.100, 0.150 and 0.200. The sintered pellets were either lapped to a thickness of either ~1mm and characterised dielectrically, or to ~2mm for the measurement of electrical d.c. conductivity properties. Samples which had been dielectrically characterised were then sawn in half and the exposed cross section mounted in epoxy resin and polished. Selected samples were examined with EPMA, and then all samples were thermally etched according to the method used by Lee and Freer [12] and examined by SEM.

RESULTS

Dielectric characterisation of the samples revealed that extension of the calcination time caused some compositions to exhibit d.c. conductivity. This behaviour occurred for both x = 0.30 and 0.35, for the y values of 0.000, 0.005, 0.010, 0.015, 0.020, 0.150 and 0.200. These will be referred to for the rest of this paper as low and high dopant level samples. The y values 0.050, 0.075 and 0.100, hereby denoted mid dopant level samples, did not exhibit this conduction behaviour. The trends of the relative permittivity and the temperature of maximum relative permittivity are reported elsewhere [13], and generally follow the pattern of results reported in other studies involving the systematic doping of SBN materials - i.e. as the dopant level increases both the maximum relative permittivity and the temperature at which it occurs fall.

The d.c. conductivity results for the samples confirmed the trend observed from the dielectric characterisation, with the low and high dopant level samples conducting more than the mid level samples. Characterisation was complicated by the thickness of the samples being too high, which required a high voltage to achieve a measurable current. This introduced the risk of heating effects in the sample, therefore the applied voltage was limited to a maximum of 750 volts.

The EPMA results showed that, as well as the parent SBN matrix, there were at least two other phases present. The analyses were complicated both by the fine nature of the other phase particles (in the order of a few microns in diameter, or less) in all but the highest dopant level samples, and by overlaps of the x-ray lines of Ba & Ti and Sr & W in the WDX spectra. Because of this, although the analyses are directly comparable with each other, in terms of absolute composition they are semi-quantitative. The approximate compositions of the other phases analysed were $(Ba_zSr_{1-z})W$ which was evident on the image as light areas, and $(Ba, Sr)_2Ti_2Nb_3$ which appeared dark (oxygen was not quantified).

It was also found that the matrix composition varied at an intra-granular level, with the Sr/Ba ratio fluctuating. This was clearly visible for the sample SBN35 with y = 0.075 (Table I), where the matrix composition varied widely from the intended composition.

Table I: Variation in composition of SBN35, x = 0.075 (values in mol. fraction), for circled areas A, B, C in Figure 1.

Area	Sr	Ba	Nb	Ti	W
A	0.429	0.571	1.886	0.068	0.031
	0.384	0.616	1.845	0.096	0.020
	0.421	0.579	1.952	0.056	0.053
B	0.329	0.671	1.967	0.058	0.052
	0.305	0.695	1.917	0.077	0.062
C	0.268	0.732	2.045	0.096	0.126
	0.285	0.715	1.991	0.101	0.113

The areas referred to in Table I relate to Figure 1: a false colour digital back scattered electron image from the EPMA of the same sample. It is evident that there are both grains and grain boundaries which exhibit enhanced average atomic number contrast, although the resolution compared with secondary electron micrographs is poor. The Sr/Ba ratio variation is most visible at the grain boundaries, indicated as C, which are Sr deficient and W rich. The analyses in Table I do not relate directly to the labeled areas, but are representative of the wide inhomogeneity of element distribution within the parent SBN at a very local level. As the dopant concentration increases so the varying composition of the matrix with regard to the dopant and Sr/Ba content becomes more obvious. The matrix of SBN30 with y = 0.200 actually has an x value of between 0.36 and 0.39, with regions of intermediate composition.

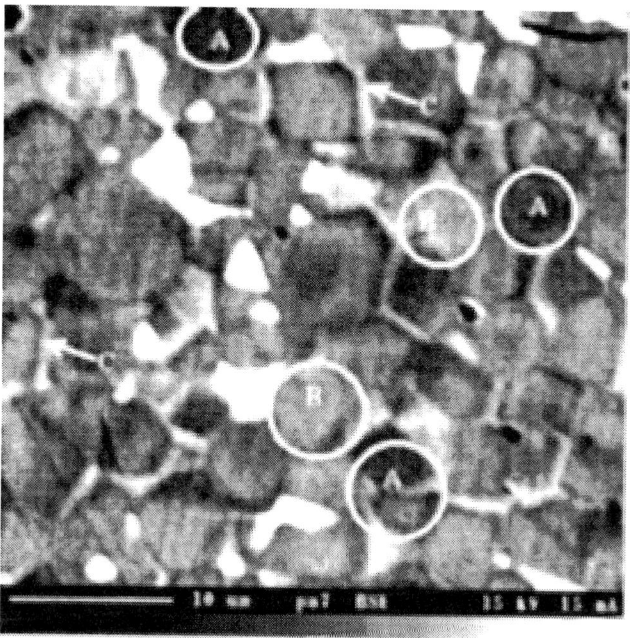

Figure 1: EPMA digital back scattered electron false colour image of SBN30 x = 0.075 (circled areas A, B, and C are referred to in Table I).

Detailed peak fitting of the XRD trace for SBN30 x = 0.200 identified the parent phase as a close match to SBN50, JCPDS file number 39-265, and two other phases of composition $Ba_3TiNb_4O_{15}$ (JCPDS 34-120) and $Ba_3Ti_5Nb_6O_{28}$ (JCPDS 37-1477) respectively. The former has a TTB structure with lattice parameters very close to those of SBN50 and the latter is monoclinic, the dielectric properties of which have been reported by Roberts et al [14]. There is also a similar monoclinic phase described by Gasperin [15], with the composition $Ba_2Ti_3Nb_4O_{18}$, and the same lattice parameters as those reported by Roberts. At this stage it is not known which phase is actually present. Due to the semi-quantitative nature of the EPMA analyses carried out, it is likely that the (Ba, $Sr)_2Ti_2Nb_3$ phase detected is actually one of these phases, with Sr substituting on the Ba sites. There were also a number of peaks which remained unattributed after this treatment. However a theoretical pattern was created[16] for $(Ba_zSr_{1-z})WO_4$ using the traces for the individual, isostructural compounds $BaWO_4$ and $SrWO_4$. This was compared to the remaining data and found to be a good match.

The phases detected in the x = 0.200 sample were also found in the other samples where dopant concentrations were high enough for XRD to detect any other phases formed. The value of z varied between 0.6 and 0.85, depending on the dopant content of the sample.

The peaks for the main SBN phase were identified by a number of methods. The first was to use the JCPDS data 39-265 for SBN50 as a guide. This method was complicated by the fact that one of the potential other phases has a crystal structure which is very similar to this. The second approach was to create theoretical patterns for SBN30 and 35 using lattice parameters reported in literature [10, 16]. These were then used as a guide, and were found to be very accurate for the experimental traces. Once the peaks in the samples had been indexed the lattice parameters were extracted from them using the computer program Crystallographica [18], and these parameters are displayed in Figures 2, 3 and 4. The results for the c-lattice parameter, Figure 2, clearly show that the general trend is as the dopant concentration increased the c-lattice parameter decreased. This parameter represents the height of the unit cell (one niobium oxygen octahedron high) a result which indicates that the dopant ions, which both have a smaller ionic radius than Nb, are occupying the Nb sites as intended. The plateau on the SBN35 graph corresponds to the samples which do not exhibit conducting behaviour: y = 0.050, 0.075 and 0.100.

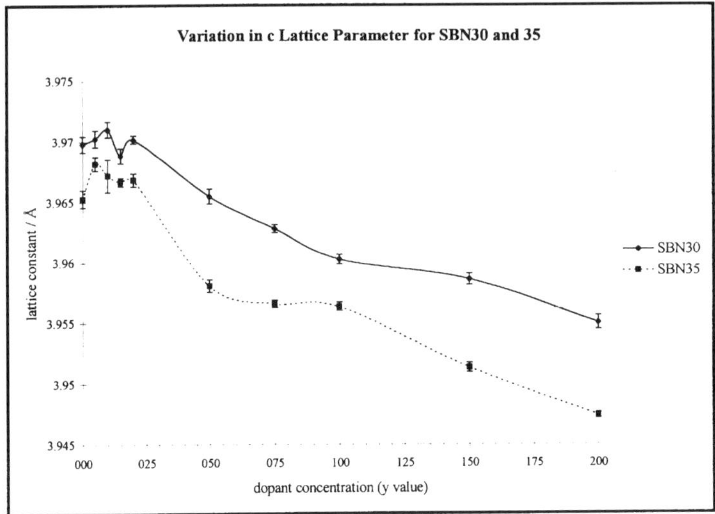

Figure 2: Variation of the c-lattice parameter with dopant concentration.

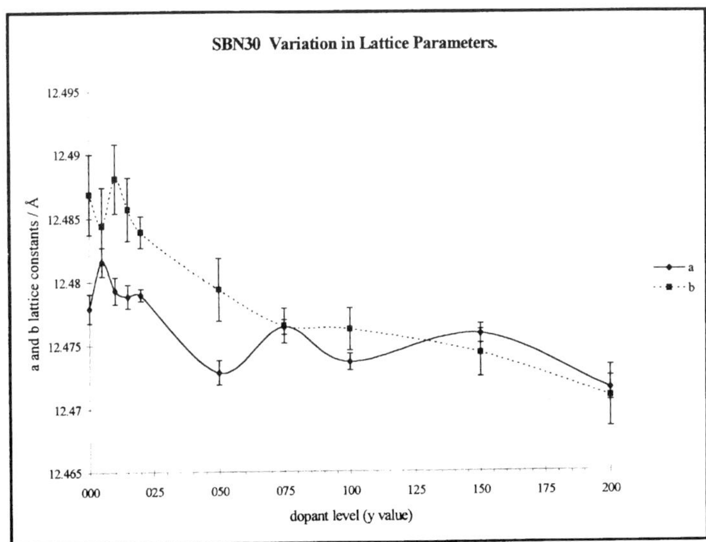

Figure 3: Variation in the a and b Lattice Parameters with dopant content for SBN30.

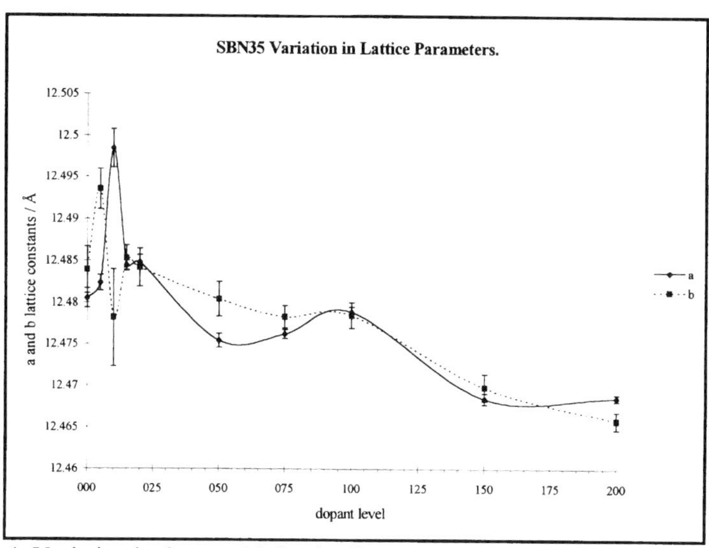

Figure 4: Variation in the a and b Lattice Parameters with dopant content for SBN35

The variations in the a and b lattice parameters, Figures 3 and 4, were less clear. Generally there was a wide deviation from the expected tetragonality of the material. The parameters for the first four compositions fluctuate wildly, although this does shadow the dielectric behaviour of the materials to a certain extent [13]. As with the c-lattice parameter there is a general decrease in the values of a and b as the dopant content increases, again attributable to the introduction of smaller ions into the unit cell. The error bars displayed on the plots are the estimated standard deviations produced by the 'Crystallographica' calculations, and they do not directly account for the difference in value of a and b. These results are also summarised in Table II; data points labeled NW are based on the lattice parameters quoted by Nishiwaki et al [17] and VD are extrapolated from the data for SBN50 to SBN75 reported by VanDamme et al [10].

Table II: Summary of refined lattice parameters for SBN30 and 35.

	NW	VD	000	005	010	015	020	050	075	100	150	200
SBN30												
a	12.49	12.477	12.478	12.482	12.479	12.479	12.479	12.473	12.476	12.474	12.476	12.472
b	12.49	12.477	12.487	12.484	12.488	12.486	12.484	12.479	12.476	12.476	12.474	12.471
c	3.97	3.946	3.970	3.970	3.971	3.969	3.970	3.965	3.963	3.960	3.959	3.955
SBN35												
a		12.472	12.481	12.482	12.498	12.484	12.485	12.475	12.476	12.479	12.468	12.469
b		12.472	12.484	12.494	12.478	12.485	12.484	12.480	12.478	12.478	12.470	12.466
c		3.942	3.965	3.968	3.967	3.967	3.967	3.958	3.957	3.956	3.951	3.947

TEM/SAED examination of SBN30 with x = 0.200 revealed the presence of two Ba, Ti, Nb phases - one phase was of high symmetry and the other low, backing up the findings of the XRD peak fitting. TEM/EDX also showed the presence of (Ba, Sr)WO_4 with a z value of 0.9. XPS analysis of the same powder suggested that the three transition elements present; Nb, Ti and W, were all in a variety of oxidation states. This result is complicated by the possibility that the process of ion beam sputter cleaning of the powder surface may itself be reducing [19].

DISCUSSION/CONCLUSIONS

In the processing of SBN30 and 35, an extension of the calcination time to 24 hours allows other phases to form as well as the expected SBN. This extension also appears to encourage d.c. conductivity behaviour, and inter- and intra-granular inhomogeneity of composition at a very discrete level both in terms of the concentration of dopant ions and the Sr/Ba ratio.

The variation of the matrix composition can perhaps be explained with reference to the work of Glass [6] who stated that of the two sites (A1 and A2) shared by the Sr and Ba ions in the unit cell, one was larger than the other; the Ba ions favour the larger site (A2) while the Sr ions will occupy either. Therefore as the Ba content increases so will the level of segregation between the two sites, with the Sr ions becoming increasingly concentrated on the smaller sites (A1). At the composition SBN33/34 the ions would be completely segregated on the separate sites and the material would be at an entropy minimum - below this value of x the Ba ions would have filled the favoured sites and therefore would have to occupy the smaller sites as well. It is postulated here that this situation is energetically unfavourable for the Ba ions, so when the situation arises whereby the material is held at a high temperature for long enough the Ba ions in the unfavourable sites start to form other phases instead. As the Ba combines with the other ions the matrix phase becomes relatively Sr rich - i.e. the x value of the matrix is effectively increased until it is above the entropy minimum and the relative instability is decreased. Thus the matrix contains not one SBN but a series, each with a different value of x. This postulate is backed up by the EPMA results, where the variation of composition is clearly visible both as average atomic number contrast on back scattered electron images, and from elemental analyses.

The variation of the lattice parameters seems to indicate that the dopant ions are entering the matrix to a certain extent, and in particular are occupying the Nb sites as intended. This conclusion can be drawn from the decrease of the c lattice parameter with increasing dopant content - as mentioned earlier the unit cell of SBN is one NbO_6 octahedron high. Thus if the Nb is replaced by a smaller ion the height of the unit cell will decrease. The variation of a and b lattice parameters from tetragonality has been reported elsewhere[1,2] although the wide variation for the low dopant compositions may be a consequence of the formation of other phases. The plateau observed for the c lattice parameter of the SBN35 compositions, which coincides with the non-conducting samples, seems to indicate that at this level of doping ($y = 0.050$) the dopant ions no longer go into solid solution in the matrix, but start to form the other observed phases. As the dopant concentration increased to a y value higher than 0.100, the dopant ions were shown by EPMA to once again migrate into the matrix and the c lattice parameter again decreased. The underlying thermodynamics which control the mechanism of formation of a solid solution or other phase are complex, and would need to be the subject of other work.

Future studies in this subject would involve the investigation of the correlation between fluctuating lattice parameters, the dielectric and conducting behaviour, and the examination by EPMA of more intermediate composition samples in order to map out when and in which regions of microstructure the other phases start to form. Experiments could also be carried out with intermediate calcination periods to investigate the kinetics of formation of the other phases.

ACKNOWLEDGMENTS

PMA would like to thank Dr. Eric Condliffe at Leeds for the EPMA analyses, and EPSRC for funding the project.

REFERENCES

1/ M.F. FRANCOMBE: 'The Relation between Structure and Ferroelectricity in Lead Barium and Barium Strontium Niobates', *Acta Cryst.*, 1960, **13**, 131-140.
2/ I.G. ISMAILZADE: 'An X-ray Study of the $BaNb_2O_6$-$CaNb_2O_6$ and $BaNb_2O_6$-$SrNb_2O_6$ Systems', *Soviet Physics - Crystallography*, 1960, **5** (2), 249-252.
3/ J.R. CARRUTHERS and M. GRASSO: 'Phase Equilibria Relations in the Ternary System BaO-SrO-Nb_2O_6', *J. Electrochem. Soc.: Solid State Science*, 1970, **117**, (11), 1426-1430.
4/ P.V. LENZO, E.G. SPENCER and A.A. BALLMAN: 'Electro-optic Coefficients of Ferroelectric Strontium Barium Niobate', *App. Phys. Letts*, 1967, **11** (1), 23-24.
5/ E. L. VENTURINI, E. G. SPENCER, P. V. LENZO and A. A. BALLMAN: 'Refractive Indices of Strontium Barium Niobate', *J. Appl. Phys.*, 1968, **39** (1), 343-344.
6/ A. M. GLASS: 'Investigation of the Electrical Properties of $Sr_{1-x}Ba_xNb_2O_6$ with Special Reference to Pyroelectric Detection', *J. Appl. Phys.*, **40** (12), 1969, 4699-4713.
7/ G. A. SMOLENSKII, V. A. ISUPOV, A. I. AGRANOVSKAYA and S. S. POPOV: 'Ferroelectrics with Diffuse Phase Transitions', *Soviet Physics - Solid State*, 1961, **2** (11), 2584-2594.
8/ A. A. BALLMAN and H. BROWN: 'The Growth and Properties of Strontium Barium Metaniobate, $Sr_{1-x}Ba_xNb_2O_6$, a Tungsten Bronze Ferroelectric', *J. Crystal Growth*, 1967, **1**, 311-314.
9/ P. B. JAMIESON, S. C. ABRAHAMS and J. L. BERNSTEIN: 'Ferroelectric Tungsten Bronze-Type Crystal Structures I Barium Strontium Niobate $Ba_{0.27}Sr_{0.75}Nb_2O_6$', *J. Chem. Phys.*, 1968, **48** (11), 5048-5057.
10/ N. S. VANDAMME, A. E. SUTHERLAND, L. JONES, K. BRIDGER and S. R. WINZER: 'Fabrication of Optically Transparent and Electro-optic Strontium Barium Niobate Ceramics', *J. Am. Ceram. Soc.*, 1991, **74** (8), 1785-1792.
11/ T. FANG, N. WU and F. SHIAU: 'Formation Mechanism of Strontium Barium Niobate Ceramic Powders', *J. Mats. Sci. Letts.*, 1994, **13**, 1746-1748.
12/ H.Y. LEE and R. FREER: 'The Mechanism of Abnormal Grain Growth in $Sr_{0.6}Ba_{0.4}Nb_2O_6$ Ceramics', *J. Appl. Phys.*, 1997, **81** (1), 376-382.
13/ P.M. ARRONDELLE, D. HIND and N.W. THOMAS, 'The Effect of Processing Conditions on the Dielectric Properties of SBN Ceramics', in press.
14/ G.L. ROBERTS, R.J. CAVA, W.F. PECK Jr and J.J. KRAJEWSKI: 'Dielectric Properties of Barium Titanium Niobates', *J. Mater. Res.*, 1996, **12** (2), 526-530.
15/ M. GASPERIN: 'Synthesis and Structure of a New Titanoniobate: Barium Tri-titanoniobate, $Ba_2Ti_3Nb_4O_{18}$', *Acta. Cryst.*, 1984, **C40**, 9-11.
16/ N.W. THOMAS: Private Communication.
17/ S. NISHIWAKI, J. TAKAHASHI and K. KODAIRA: 'Effect of Additives on Microstructure Development and Ferroelectric Properties of $Sr_{0.3}Ba_{0.7}Nb_2O_6$ Ceramics', *Jpn. J. Appl. Phys.*, 1994, **33** (1, 9B), 5477-5481.
18/ Crystallographica Version 1.10 (August 1996): Oxford Cryosystems, Long Hanborough, Oxford.
19/ T. CHOUDRY, S.O. SAIED, J.L. SULLIVAN and A.M. ABBOT: 'Reduction of Oxides of Iron, Cobalt, Titanium and Niobium by Low Energy Ion Bombardment', *J. Phys. D. : Appl. Phys.*, 1989, **22**, 1185-1195.

ELASTIC MODULUS, MODE I (OPENING) FRACTURE TOUGHNESS AND THE CRITICAL DEFECT SIZE OF Bi(Pb)SrCaCuO EXTRUDED RODS.

O.O. Oduleye, S.J. Penn and N.McN. Alford
South Bank University, 103 Borough Road, London SE1 0AA
Email oduleyoo@sbu.ac.uk
s.penn@sbu.ac.uk
alfordn@sbu.ac.uk

ABSTRACT

Experiments have been carried out to determine the elastic modulus in bending, E_B, the mode1 (opening) fracture toughness, K_{1c} and the critical flaw size, a_0 of 1 mm extruded (Bi,Pb)SrCaCuO rods. These properties are crucial in developing a clearer idea of the behaviour of BSCCO superconductors in service and are a pre-requisite for modelling using for example finite element techniques. The experiments show the effect of pore volume on these mechanical properties. It is found that the critical defect length is BSCCO is surprisingly large and this has important implications on the processing of the materials and to the limits of application.

INTRODUCTION

Ag/BSCCO-2223 tapes can carry a critical current density, J_c, of 60 kA cm^{-2} @ 77K 0T[1] in pressed tapes of a few centimetres in length. Such a high J_c has, so far, not been achieved in longer samples. Longer samples are characterised by inhomogeneities in the critical current distribution and in the mechanical properties. In long samples the local critical current is influenced by the mechanical properties of that section. Of particular interest are the core density, ρ, and the associated volume fraction porosity, p; the elastic modulus, E; and fracture mechanics parameters such as the mode I (opening) fracture toughness, K_{Ic}, and the critical flaw size, a_0. E, K_{Ic} and a_0 are found to be dependent on volume fraction porosity.

One method of preparation of Ag/BSCCO tapes is the powder-in-tube method (PIT). The PIT involves several mechanical deformation and heat treatment processes. The final aim beig to achieve optimum a-b plane alignment and densification of BSCCO-2223 plates. This is necessary for a high critical current density in the tapes.

The Ag/BSCCO system is a composite of ductile and brittle materials. The thermal expansion mismatch between the two materials can encourage crack formation and propagation in the ceramic core when the composite is subjected to cyclic thermal stresses. Furthermore, superconductors will be subjected to various stresses under working conditions. In order to carry out a complete stress and fracture mechanics analysis of the Ag/BSCCO composite tapes under working conditions, it is necessary to determine the mechanical properties. In this paper, we present our results of the elastic modulus, E, and the mode I fracture toughness, K_{Ic}.

EXPERIMENT

Cylindrical 1 mm BSCCO rods were prepared from BiPbSrCaCuO starting powders (Merck). The powder was mixed in a high shear mixer using polymers and organic solvents until a "dough-like" consistency was achieved. The plastic material was then extruded in a ram extruder forming long lengths of material either 1 or 3mm in diameter depending on the die size[2]. Five batches of 1 mm rods were sintered at various temperatures as shown in table I.

Batch	Temp. (°C)	final density (average) (g cm^{-3})
A	855	4.25
B	850	5.17
C	840	4.07
D	830	4.00
E	810	3.55

Table I : Sintering temperature and final density.

The Young's modulus in bending, E_B, was determined by centre point loading using the expression:

$$E_B = \frac{4PL^3}{3\pi D^4 y} \tag{1}$$

Where P is the ultimate load in the rod, L is the span between supports, D is the diameter of the rod and y the deflection. In order to avoid measuring shear components, span to diameter ratios greater than 50 were used.

Fracture toughness (K_{Ic}) testing was carried out by means of centre loading point (3 point bend test) on single edge notched round bars in bending[3, 4, 5]. The fracture toughness, K_{Ic}, and the critical crack length, a_0, were determined for rods sintered at 850° C for 2 hours to 63.1% of theoretical density. Before sintering, the rods were prepared with notch lengths, a, in the range 120-240 μm using a low speed saw with a 150 μm diamond blade. The average diameter of the sintered rods was 0.833 mm. The critical stress intensity factor in mode I was determined using equations 2 and 3, 4 and 5 and 4 and 6:

$$K_{1c} = \frac{32M}{D^3}\sqrt{\frac{a}{\pi}}\left(Y\left(\frac{a}{D}\right)\right) \tag{2}$$

where

$$Y\left(\frac{a}{D}\right) = 1.04 - 3.64\left(\frac{a}{D}\right) + 16.86\left(\frac{a}{D}\right)^2 - 32.59\left(\frac{a}{D}\right)^3 + 28.41\left(\frac{a}{D}\right)^4 \tag{3}$$

and M is the maximum bending moment in the rod.

$$K_{1c} = 0.25\left(\frac{L}{D}\right)Y'\frac{P}{D^{1.5}} \tag{4}$$

where L/D is the support span/diameter ratio and Y' is a correction factor dependent on specimen geometry and type of loading.

Y' for single edge notched round bars in bending is given by Bush [5] as:

$$Y' = 12.7527\frac{(a/D)^{0.5}}{(1-a/D)^{0.25}} \tag{5}$$

and by Ouchterlony [3] as:

$$Y' = 12.7527\frac{(a/D)^{0.5}\left[1+19.646(a/D)^{4.5}\right]^{0.5}}{(1-a/D)^{0.25}} \tag{6}$$

The measured K_{1c} may only be regarded as valid if the specimens are notch sensitive. This means that the presence of a notch should reduce the measured stress over and above that which would be expected taking the net section stress. The notch sensitivity, N, is defined as[6, 7].

$$N = \frac{\sigma_n}{\sigma_f} \tag{7}$$

Where σ_n is the net section failure stress for the fracture of a notched specimen and σ_f is the fracture stress of an unnotched specimen. σ_f is calculated using the expression:

$$\sigma_f = \frac{8PL}{\pi D^3} \tag{8}$$

For a notched round bar of circular cross-section as shown in figure 1 below, σ_n is calculated using the expression:

$$\sigma_n = \frac{P_n L y_{max}}{4I_{GG}} \tag{9}$$

where I_{GG} is the second moment of area about and axis passing through the centroid of the section and parallel to the x-axis and is calculated using the expression:

$$I_{GG} = \frac{\pi r^4}{8} - \frac{r^4 \sin^{-1}\frac{p}{r}}{4} - \frac{p}{8r^4}\sqrt{r^2-p^2}\left(r^2-2p^2\right) - \left[\frac{4r^6}{9}\left(1-\frac{p^2}{r^2}\right)^3\right]\left[\frac{\pi r^2}{2} - \sin^{-1}\frac{p}{r} - p\sqrt{r^2-p^2}\right]^{-1} \quad (10)$$

Figure1 is a schematic representation of the cross-sections of the rods. The origin of the co-ordinate axes is at *O*. The distance $\bar{y}$ measured from the origin along the y-axis represents the centroid of the section. The distance *p* is measured from the origin along the y-axis to the bottom surface of the 'unnotched' section. The dashed arc represents the notched section, *r* is the radius of the rod and *D* its diameter *a* is the notch depth.

$$\bar{y} = \frac{2r^3}{3}\left[1-\frac{p^2}{r^2}\right]^{\frac{3}{2}}\left[\frac{\pi r^2}{2} - \sin^{-1}\frac{p}{r} - p\sqrt{r^2-p^2}\right]^{-1} \quad (11)$$

and

$$y_{max} = r + \bar{y} - a \quad (12)$$

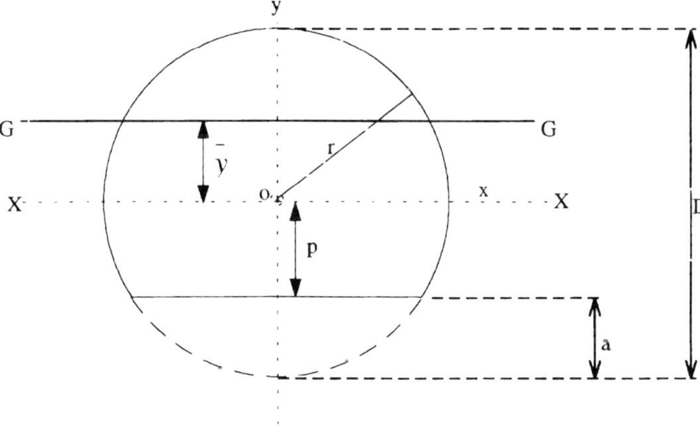

Figure 1: Schematic representation of the single edge notched round bar specimen.

Density of the sintered rods was determined by mass and dimensions and assuming a theoretical maximum density of 6.5 g cm^{-3} for BSCCO 2223 [8].

RESULTS

Young's Modulus, E_B

Figure 2 below shows the Young's modulus of the rods as a function of volume fraction porosity, p. The highest measured value of Young's modulus is 78.63 GPa for rods sintered at 850 °C for 2 hours to 80% of theoretical density.

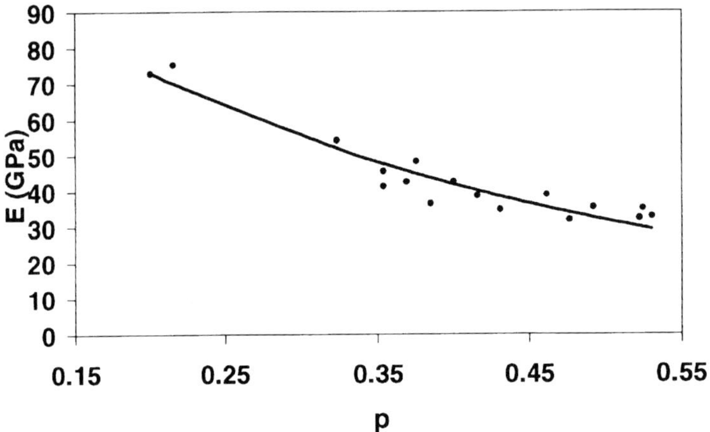

Figure 2: Variation of elastic modulus with volume fraction porosity, p: solid line $E= 127e^{-2.77p}$.

The variation of Young's modulus can be fitted to the expression:

$$E_B = E_0 e^{-np} \tag{13}$$

The extrapolated elastic modulus at zero porosity, E_0, is found to be 127 GPa, and n in equation (13) is found to be 2.77. The value of E_0 is in agreement with the results of Alford et al[9] for results on $YBa_2Cu_3O_{7-\delta}$ and $La_{1.8}Sr_{0.2}CuO_4$, but higher than the values used by Ochiai et al[10] and about half the value given by Mackrodt for atomistic calculations of the elastic properties of BSSCO (2223) where E is calculated to be 230 GPa [11].

Mode 1 Fracture Toughness, K_{1c} (Critical stress intensity factor)

The rods are most notch sensitive when the relative notch depth is in the range 0.2-

0.35 corresponding to N in the range 0.15-0.25. Table II below shows K_{1c} and calculated from eq. 2 and 3, 4 and 6, and 5 and 6. Comparing the results in table II, it can be seen that the K_{1c} values calculated using the three sets of equations are similar. Taking the values from the last column of table II, the fracture toughness of the rods lies in the range 0.57-0.88 $MPa\sqrt{m}$ with an average of 0.67 $MPa\sqrt{m}$ and standard deviation of 0.1. This value is lower than those from measurements on YBCO by Fujimoto et al[12] and measurements on BSCCO by Goretta et al[13], Martin et al[14] and Chu et al[15] but very similar to measurements on BSCCO by Lewis [16].

notch length (mm)	diameter (mm)	Y'(eq. 5)	Y'(eq. 6)	Y (eq. 3)	K_{1C}(eq. 4&5) ($MPam^{0.5}$)	K_{1C}(eq. 4&6) ($MPam^{0.5}$)	k_{1C} (eq. 2&3) ($MPam^{0.5}$)
0.00	0.76	0.00	0.00	1.04			
0.21	0.85	6.64	6.75	0.78	0.61	0.62	0.64
0.21	0.83	6.76	6.88	0.79	0.84	0.85	0.88
0.22	0.85	6.87	7.01	0.79	0.54	0.55	0.57
0.22	0.88	6.76	6.89	0.79	0.65	0.66	0.68
0.17	0.86	5.86	5.89	0.77	0.64	0.64	0.67
0.21	0.84	6.76	6.89	0.79	0.56	0.57	0.59
0.12	0.83	4.95	4.96	0.78	0.64	0.64	0.69

Table II : A comparison of K_{1c} using equations (2) and (3), (4) and (5), and (4)and (6)

The critical crack length, a_0, can be calculated from equation (2) as:

$$a_0 = \left(\frac{K_{1c}}{\sigma_m \sqrt{\pi} Y\left(a/D \right)} \right)^2 \tag{14}$$

where σ_m has been substituted for $32M/\pi D^3$. From column 5 in table II it is seen that Y(a/D) is approximately 0.8. The highest measured flexural strength in our experiments is 89 MPa. Substituting these values into equation (14) yields a_0 = 28.1 μm.

The Effects of Porosity on K_{1c} and a_0

The strain energy release rate, *G*, and the stress intensity factor, *K*, are equivalent[17]. In mode I fracture they are related by equation the expression :

$$K_I^2 = GE\alpha \tag{15}$$

where α= 1 for plane stress and $\alpha = 1/(1-\nu^2)$ for plane strain and ν is Poisson's ratio and *E* and *G* are the elastic modulus and strain energy release rate measured at zero porosity.

The results above show that E varies with porosity and may be fitted to an equation of the form shown in equation (13). The fracture energy, R, of ceramics varies with volume fraction porosity, p, as[18].

$$R = R_0 e^{-bp} \tag{16}$$

Equation (15) can then be written as

$$K_1 = \sqrt{E_0 G_0 \alpha e^{\frac{-2.77}{2}} e^{\frac{-bp}{2}}} \tag{17}$$

Where we have assumed that volume porosity dependence of G is similar to equation (16).

In equation (17) the change in elastic modulus from a fully dense material to an 80% dense rod alone accounts for a 30% reduction in $K_{Ic.}$
Taking the above into consideration, K_{Ic} and a_0 for fully dense BSCCO is estimated to be 1.1 $MPa\sqrt{m}$ and 76.0 μm respectively.

CONCLUSIONS

The results of experiments to determine the elastic modulus, E_B, mode 1 (opening) fracture toughness, K_{1c} and the critical flaw size of 1 mm BSCCO rods have been presented. The results for the elastic modulus show a close correlation with the pore volume. The Young's modulus was found to be approximately 130 GPa on extrapolation to zero porosity. The fracture toughness has been measured on cylindrical edge notched beams and a mean value of 0.67 $MPa\sqrt{m}$ was measured for 1 mm diameter rods. Using these values, the critical flaw size was found 28 μm in the rods. The zero porosity value for K_{Ic} is estimated to be 1.1 MPa√m, this would yield a flaw size of 76 μm.

ACKNOWLEDGEMENTS

BICC Cables kindly supplied samples of Ag/BSCCO. The BSCCO green body rods were prepared by Geoffrey Dolman, of Birmingham University. This project is funded by the Engineering and Physical Sciences Research Council

REFERENCES

1 Z. HAN, P.SKOV-HANSEN and T.FRELTOFT: 'Topical Review: The Mechanical Deformation of Superconducting BiSrCaCuO/Ag Composites', *Supercond. Sci. and Technol.*, 1997 **10** 371

2 N.McN.ALFORD, J.D.BIRCHALL and K.KENDALL: 'High Strength Ceramics Through Coloidal Control to Remove Defects', *Nature*, 1987 **330** (6143) 51

3 B.I.G.BARR and E.D.B.HASSO: 'Fracture Toughness Testing by Means of the SECRBB Test Specimen', *Int. J. of Cement Composites and Lightweight Concrete*, 1986 **8** (1) 3-9

4 O.E.K.DAOUD and D.J.CARTWRIGHT: 'Strain Energy Release Rates for a Straight Fronted Edge Crack in a Circular Bar Subject to Bending', *Eng'g Fracture Mechanics*, 1984 **19** (4) 701-707

5 A.J.BUSH: 'Experimentally Determined Stress-intensity Factors for Single-edge-crack Round Bars Loaded in Bending', *Experimental Mechanics*, 1976 **16** (7) 249-257

6 E.PETERSSON : *Crack Growth and Development In Plain Concrete and Similar Materials*, Lund Institute of Technology, Sweden, 1981.

7 N.McN.ALFORD, G.W.GROVES and D.D.DOUBLE: 'Physical Properties of High Strength Cement Pastes', *Cement and Concrete Research*, 1982 **12** 349-358

8 M.POLAK, W.ZHANG, A.POLYANSKII, A.PASHITSKI, E.E.HELLSTROM and D.C.LARBALESTIER: ' The Effect of the Maximum Processing Temperature on the Microstructure and Electrical Properties of Melt Processed Ag-sheathed BiSrCaCuO Tape', *IEEE Trans. Appl. Superconduct.*, 1997 **7** 1537

9 N.MCN.ALFORD, J.D.BIRCHALL, W.J.CLEGG, M.A.HARMER, K.KENDALL, D.J.EAGLESHAM, C.J.HUMPHREYS and D.R.JONES: 'Processing and Properties of High T_c Superconductors', *British Ceramic Proceedings*, 1988 (40) 149-160

10 S.OCHIAI , K.HAYASHI and K.OSAMURA: 'Influence Of Thermal Cycling On The Critical Current Of Superconducting Silver-sheathed High T_C Oxide Wires', *Cryogenics*, 1991 **31** 954.

11 W.C.MACKRODT: 'Calculated Lattice Structure, Stability and Properties of the series $Bi_2X_2CuO_6$ (X=Ca, Sr, Ba), $Bi_2X_2YCu_2O_8$ (X=Ca, Sr, Ba; Y= Mg, Ca, Sr, Ba) and $Bi_2X_2Y_2Cu_3O_{10}$ (X=Ca, Sr, Ba; Y= Ba, Sr, Ca, Mg), *Superercond. Sci. and Technol.*, 1989 **1** 343

12 H.FUJIMOTO, M.MURAKAMI, T.OYAMA, Y.SHIOHARA N.KOSHIZUKA and S.TANAKA: 'Fracture Toughness of YBaCuO Prepared by MPMG Process', *Jpn. J. Appl. Phys.*, 1990 **29** (10) L1793

13 K.C.GORETTA, M.E.LOOMANS, L.J.MARTIN, J.JOO, R.B.POEPPEL and N.CHEN: 'Fracture Of Dense Textured $Bi_2Sr_2CaCu_2O_X$ ',*Supercond. Sci. and Technol.*, 1993 **6** 282.

14 L.J.MARTIN, K.C.GORETTA, J. JOO, J.P.SINGH, S.R.OLSON,

S.WASYLENKO, R.B.POEPPEL and N.CHEN: 'Mechanical Properties Of BiSrCaCuO/Ag Superconductors', *Materials Lett.*, 1993 **17** 232

15 C.Y.CHU, J.L.ROUTBORT, N.CHEN, A.C.BLONDO, D.S.KUPPERMAN and K.C.GORETTA: 'Fracture Of Dense Textured $Bi_2Sr_2CaCu_2O_X$', *Supercond. Sci. and Technol.*, 1992 **5** 306

16 G.LEWIS: 'Fracture Toughness Of Two Cuprate-based High-transition Temperature Ceramic Superconducting Materials', *J. Mat. Sci. Lett.*, 1992 **11** 321

17 B.LAWN: *Fracture Of Brittle Solids*, Cambridge University Press, Cambridge, 1993

18 R.W.RICE, S.W.FREIMAN, R.C.POHANKA, J.J.MECHOLSKY and C.C.WU: *Fracture Mechanics of Ceramics*, Plenum, London, 1978 **4** 849

19 S.SALIB, M.MIRONOVA, C.VIPULANANDAN and K.SALAMA: 'Mechanical Properties and TEM Studies on BPSCCO-2223-Ag Composite Tapes', *Supercond. Sci. and Technol.*, 1996 **9** 1071

20 A.S.WAGH, J.PSINGH and R.B.POEPPEL: 'Dependence Of Ceramic Fracture Properties On Porosity', *J. Mat. Sci.*, 1993 **28** 3589

21 J.TIROSH, E.ALTUS and Y.YIFRACH: 'A New Method For Evaluating The Fracture Toughness Of Brittle Materials', *Int. Of Fracture*, 1992 **58** 211

EFFECT OF PROCESSING ON THE DIELECTRIC PROPERTIES OF CERAMIC DIELECTRIC RESONATOR MATERIALS

N McN ALFORD, S J PENN, A TEMPLETON, X WANG AND S.J.WEBB
South Bank University, 103 Borough Road, London SE1 0AA, UK, alfordn@sbu.ac.uk

ABSTRACT

The world market for ceramic dielectric resonators is around 0.8BN Ecu. A major use is in cellular communications where there is explosive growth (38million subscribers in 1997 in the US alone). The massive increase in demand has caused severe congestion in the bands allocated to mobile communications (0.9 and 1.8GHz). Far better pre-selection filters are required to handle the increasing volume and one low cost and realistic option is to improve the properties of dielectric resonators by attention to processing. Certain dielectric oxide single crystals display very low loss at microwave frequencies. On cooling the loss is generally observed to drop and Q's of sapphire at 10GHz exceed 10^6 at low temperatures of around 10K. However, single crystals are expensive and the purpose of this research is to explore inexpensive, sintered polycrystalline alternatives. By very careful attention to purity, processing and microstructure Q values approaching those of single crystals have been achieved and in alumina and titania we have achieved the highest Q values ever reported in these materials at microwave frequencies. In this paper Al_2O_3 and TiO_2 have been studied as a model materials for dielectric loss and have given important insights into the role of processing as a means of optimising dielectric properties.

INTRODUCTION

The requirement for miniature low loss microwave filters has led to the dielectric loading of cavity resonators to form dielectric resonators. The miniaturisation factor can be expressed as

$$D \approx \frac{\lambda_0}{\sqrt{\varepsilon_r}}$$

where D is the diameter of the dielectric resonator, λ_o is the free space wavelength at the resonant frequency and ε_r is the dielectric constant of the dielectric. There are three key parameters for the dielectric materials in filter applications: the dielectric constant (ε_r), the quality factor ($Q = 1/\tan\delta$) and the temperature coefficient of resonant frequency (τ_{cf}). The larger the dielectric constant, the smaller the filter. The higher the Q the lower the insertion loss and steeper the cut-off. A τ_{cf} close to zero ($< \pm$ 10 ppm/K) is required for stability against ambient temperature change. The focus of this work has been to improve the properties of existing dielectrics through careful processing. Two materials have been studied here, Alumina (ε_r = 10) and TiO_2 (ε_r = 100). These two materials were chosen as model materials as part of a programme exploring the causes of microwave loss in dielectric ceramics. The reason for studying alumina is to investigate the causes of dielectric loss. Commercial materials, such as Zr-Sn-Ti-O and Ba-Mg-Ta-O, are complex multiphase materials which are very sensitive to the processing conditions. As a result it is difficult to separate the effect of different sources of loss. Alumina, however, is a single phase material which is available in a very pure form.

Because many microwave dielectric materials contain Ti eg Ba-Ti-O, Ba-Nd-Ti-O, (Ba-RE-Ti-O, RE=Rare Earth)[1], Zr-Sn-Ti-O etc it is essential to understand the role of the titanium, particularly as it can exist in mixed valence states, and the role of oxygen and its influence on the dielectric loss. Ti - containing ceramic materials have found uses as dielectric resonator materials[1]. In order to sinter the material, temperatures between 1000°C-1500°C are required. This can cause the TiO_2 to lose oxygen producing a high concentration of oxygen vacancies. One of the main issues affecting the loss tangent is the non-stoichiometry believed to be the result of oxygen vacancies.

EXPERIMENTAL

Alumina and titania discs were produced from commercial high purity powders. The discs were produced by pressing appropriate powder in 13 mm diameter stainless steel dies at a pressure of 100 MPa. The resulting discs were sintered in air at between 1000 °C and 1600 °C for between 5 and 1800 minutes.

The Q and was measured by a resonant cavity method using the $TE_{01\delta}$ mode. The sample is placed in a copper cavity on a 4 mm high low loss quartz spacer. The surface resistance of the copper has been calculated from the Q of the TE_{011} resonance of the empty cavity to allow the results to be corrected for the loss due to the cavity walls[2]. The sintered samples were approximately 10 mm diameter, 4 mm thick discs. The $TE_{01\delta}$ mode was examined using a HP8720C vector network analyser with 1 Hz resolution. The dielectric constant, ε_r was measured using the same equipment but without the quartz spacer and with the cavity height close to the size of the sample. The measurement cavity is attached to a closed cycle cooler to allow measurements to be made from 10 K to 300 K. The discs were measured in the "as-fired" condition, i.e. without surface preparation such as polishing.

ALUMINA

Alumina (Al_2O_3) is the polycrystalline equivalent of sapphire. It is a useful dielectric in it own right and is capable of having a very high Q [3,4]. However, with its low dielectric constant its use is limited to high frequency applications (ideally satellite communications > 10GHz) due to size considerations.

The porosity, purity and grain size of the sintered sample all influence the Q [4]. The influence of the grain size can be seen in fig 1. For this study all the samples were near fully dense and only the grain size was altered. Small amounts of porosity (< 3%) significantly reduce the Q as seen in figure 2 .

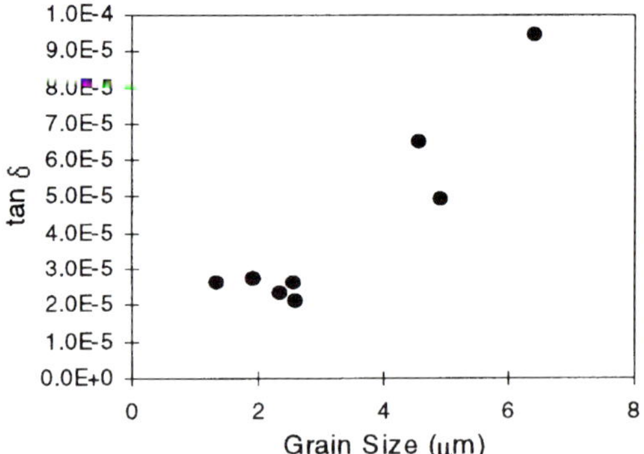

Figure 1 Influence of grain size on the tan δ of sintered alumina.

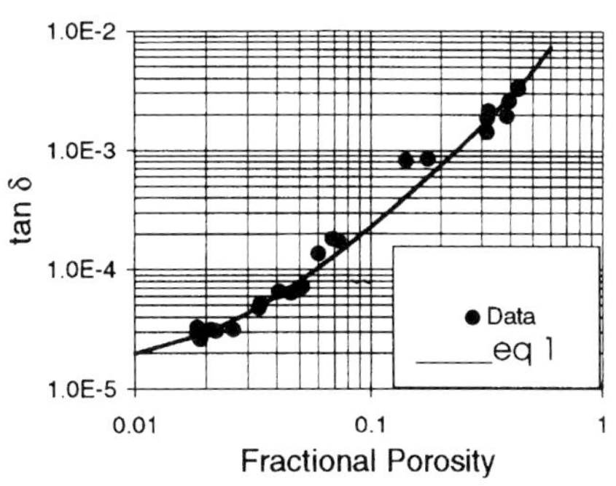

Figure 2 Influence of porosity on the tan δ of alumina.

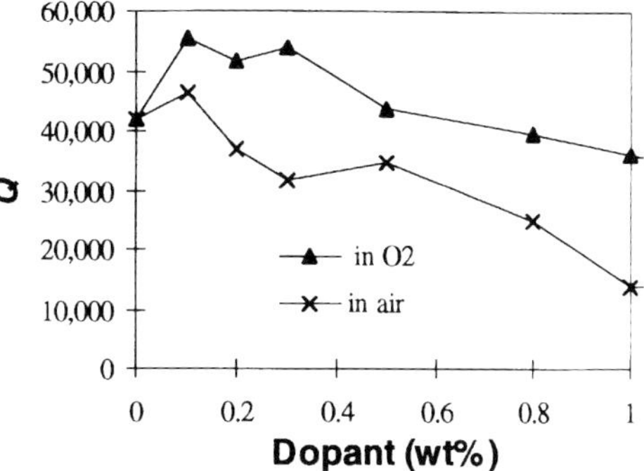

Figure 3 Effect of TiO_2 doping in alumina and influence of sintering atmosphere.

The effect of porosity be expressed in terms of eq 1.

$$\tan\delta = (1-P)\tan\delta_0 + A'P\left(\frac{P}{1-P}\right)^{\frac{2}{3}} \qquad (1)$$

Where $\tan\delta_0 = 1.565 \times 10^{-5}$ and $A' = 9.277 \times 10^{-3}$ and P is the pore volume.
It was found[3] that doping alumina with TiO_2 increased the Q by over 30%, a most unexpected result. The room temperature Q at 9 GHz reaches a maximum of 55,000 ($\tan\delta = 1.82 \times 10^{-5}$) at a dopant level of 0.1% -0.3% when sintered in flowing oxygen as shown in Fig 3. The precise reason for this is unclear. The TiO_2 used has a far higher dielectric loss than alumina ($Q \approx 2000$ at 300K, 3 GHz) but does aid sintering, reducing the temperature required and producing a fine grained dense microstructure. For higher concentrations of TiO_2 dopant the Q is observed to fall and this is associated with the dielectric losses in the TiO_2 (or calcium titanate) rather than grain size differences. At present the modification of microstructure is the preferred explanation for the enhancement in Q. Also seen in Fig 3 is the significant effect of sintering in oxygen in reducing the dielectric loss.

The tan δ ($=1/Q$) of alumina decreases approximately linearly with temperature. Theory predicts that the tan δ of sapphire should follow a T^5 dependence[5]. This appears to be true for temperatures greater than about 100 K. However, at low temperature the temperature dependence is much weaker, similar to that of the polycrystalline alumina. This is because at low temperature the dielectric loss is dominated by extrinsic effects such as impurities and defects. It is these factors that limit the Q of polycrystalline materials.

TITANIA

TiO_2 itself can have a high Q at a high dielectric constant but as a dielectric resonator material it is unsatisfactory because of the magnitude of the temperature coefficient of the resonant frequency (*Tcf*) which is influenced by changes in dimension and changes in the dielectric constant due to thermal effects. O'Bryan[6,7,8] demonstrated that Ti bearing ceramics, notably barium titanate could be prepared with compositions in which the *Tcf* could be controlled but at the expense of reduced ε_r or Q. For comparison a single crystal of rutile supplied by ESCETE (The Netherlands) was also measured and this was found to have a tan δ of 5.5 x10^{-5}

at 3.4 GHz and at room temperature. Low temperature measurements were performed by placing the cavity on a closed cycle cryo-cooler.

Dielectric Constant.

Fig 1 shows the results for the dielectric constant of the TiO_2 sintered in air. The dielectric constant behaves as expected in a dielectric containing porosity It was found to fit equation 2 well: where ε_1 is the dielectric constant of pure TiO_2, P is the volume of pores within the sample and ε is the dielectric constant of the sintered sample.

$$\varepsilon = \varepsilon_1[1 - \frac{3(\varepsilon_1 - 1)}{2\varepsilon_1 + 1} P] \qquad (2)$$

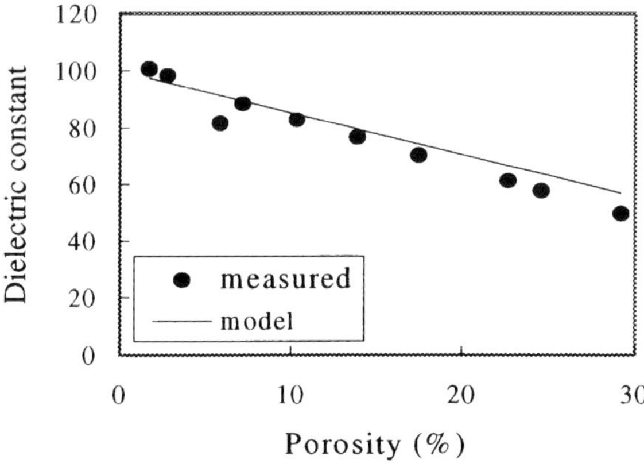

Figure. 4 Influence of pore volume on the dielectric constant of sintered TiO_2.

Dielectric Loss

In fig 5 the unloaded-Q (=1/tan δ) of the sintered TiO_2 is plotted as a function of sample porosity. An interesting observation is made here. The Q rises and then falls as densification proceeds One possible explanation for this is oxygen deficiency. At a temperature of 1400°C the sample is near fully dense and on cooling there is very little free surface capable of absorbing oxygen. At 95% of theoretical density there is no interconnected porosity and as a consequence the Q is reduced further due to the presence of the porosity. At 94% of theoretical density the Q increases sharply and this can be explained by the fact that there is now a interconnected network of pores reaching more of the surface of the TiO_2. The material is better oxygenated and the Q increases. Below a density of 94% the Q then drops due to the presence of increasing amounts of porosity. "Coring" (a well defined dark region in the centre of the sample) was observed in all the dense (and impermeable) samples, strongly suggestive of a lower oxygen content and also suggesting that oxygen had diffused only a short distance into the surface of the sample. Such effects have been observed in barium titanate[9]. Wavelength dispersive electron probe microanalysis however, did not reveal any significant differences between the darker region and elsewhere. The accuracy of this technique is ±0.2% suggesting that the degree of reduction is below this level. The results suggest that there is only very limited reduction of the TiO_2 but that this is sufficient to cause a severe deterioration in the dielectric loss. Negas et al[9] found that in $BaTi_4O_9$ and $Ba_2Ti_9O_{20}$ based ceramics low concentrations of lattice defects caused by reduction degraded the *Q* factor by almost 100%.

The addition of yttria stabilised zirconia

On the assumption that the drop in *Q* at high density was indeed an oxygenation problem, 3 mole % yttria stabilised zirconia (YSZ) powder (Mandoval HSY3) was added to the TiO_2 powder. YSZ is well known for its oxygen ion conduction properties. If YSZ were added to the TiO_2 powder at the 5-10wt% level, the reasoning was that such a level is approximately the percolation threshold and thus the YSZ would effectively act as a conduit for oxygen during cooling in the furnace. The diffusion of oxygen[10] in dense TiO_2 at 1400°C is $2.5 \times 10^{-11} cm^2s^{-1}$ whereas in calcia stabilised zirconia (which we will assume is approximately similar to YSZ) it is $2 \times 10^{-6} cm^2s^{-1}$. Thus in a 10mm diameter sample of thickness 4mm it would be impossible to oxygenate the sample in a reasonable time with a diffusion length of 2mm (approximately 28 years) but with a diffusion length on the order of the grain size eg 5 microns, a short time would be required on the order of 0.4 hours for TiO_2 and only 0.035 seconds for partially stabilised zirconia.

Figures 5 and 6 show that adding YSZ does indeed have a significant effect and the *Q* is enhanced even at very low porosity. However, the original premise that the percolation threshold of approximately 5% porosity would be the appropriate volume fraction of zirconia is clearly incorrect. The enhancement in *Q* is observed at very low dopant concentrations of YSZ. Samples of TiO_2 doped with unstabilised ZrO_2, which is not an oxygen ion conductor, were made for comparison. It was found that the *Q* was not improved. In further experiments additions of aluminium oxide at the 100ppm and 500ppm levels, manganese oxide and yttrium oxide at the same levels also caused an enhancement in the Q suggesting that the addition . The issue of oxygenation and reduction in TiO_2 is highly complex. Doping with manganese which can exist invariable valency states is believed[9] to result in the reaction $Mn^{3+} + Ti^{3+} \rightarrow Mn^{2+} + Ti^{4+}$ but why doping with Al^{3+} or Y^{3+} should cause an increase in the Q is not understood at present and warrants further investigation. The fact that YSZ is an oxygen ion conductor appears, at present, to be fortuitous.

Low temperature measurements on TiO_2.

The mechanisms and the physics of loss can sometimes be elucidated by examining the loss as a function of temperature. Theories of dielectric loss based on phonon models have been suggested by Gurevich et al [11]. These models show that the dielectric loss of TiO_2 single crystals should observe a T^3 dependence. In reality very few single crystals observe the

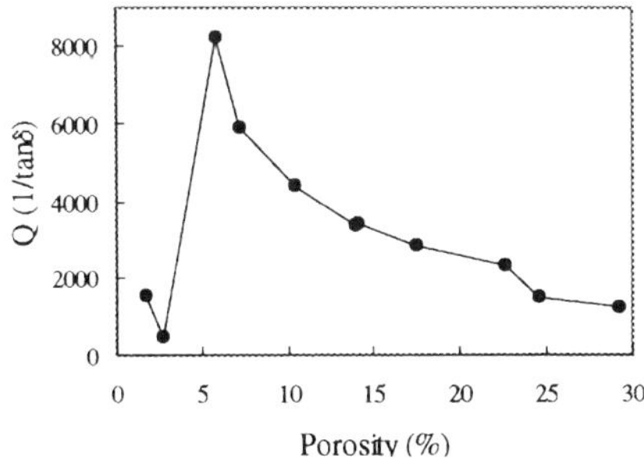

Fig.5 *Q* (=1/tanδ) as a function of the porosity at 3GHz

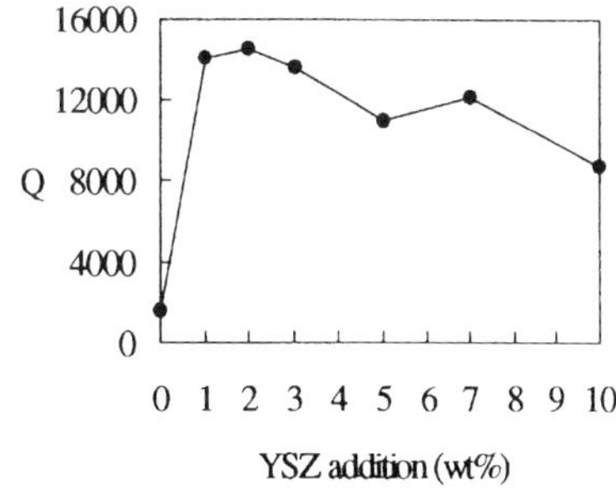

Fig.6 *Q* as a function of addition o YSZ at 3GHz

predicted temperature dependence suggested by Gurevich et al [11] and very high quality sintered alumina observes a T dependence rather than a predicted T^5 dependence. The polycrystalline TiO_2 observes a T dependence rather than a T^3 dependence. This suggests that the symmetry rules for single crystals which determine the temperature dependence are completely absent in sintered ceramics and this is hardly surprising. However, there are other features of the temperature dependence of the dielectric loss which also give important clues concerning purity, and in particular so far as Ti containing ceramics are concerned, the oxygen stoichiometry.

The dielectric loss and the dielectric constant were measured as a function of temperature. Here it is seen that this type of measurement is very sensitive to the nature of the original powder used to manufacture the discs. It gives us important information on how to tailor materials in order to obtain low dielectric loss. In much the same way as impurities in Al_2O_3 were found to influence the tan δ significantly, it now appears that the same is true for TiO_2. Fig 7 shows the tan δ vs. temperature plot for dense discs of the TiO_2 and it is clear that there are differences in the behaviour. The magnitude of the dielectric loss is different at room temperature and in the undoped material the loss is fairly constant until a temperature of approximately 100K where there is a well defined change of slope in the tan δ vs. T curve. At this temperature the tan δ drops sharply. In the doped material the loss is lower throughout most of the temperature range but again falls sharply at a temperature of around 20K. It is noted that the YSZ doped TiO_2 possesses a higher loss at 20K than the undoped material which achieves very low tan δ at 20K and below.

This is rather unusual behaviour. In contrast, when single crystal Al_2O_3 is measured at low temperatures, there appear to be defects which cannot be made inoperative at low temperature[5] and lead to a substantial residual loss at low temperatures. For comparison the behaviour the single crystal rutile is also shown. Here it is seen that there is no such inflection at low temperature and the loss drops steadily.

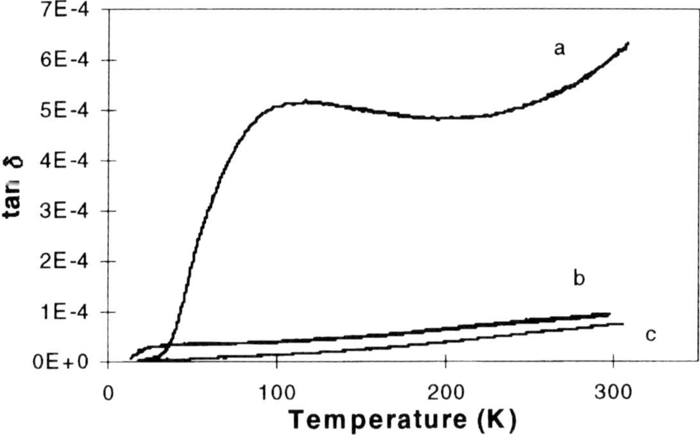

Figure 7 Tan δ as a function of temperature for TiO_2 at 3 GHz (a) undoped, (b) doped with 10 wt% YSZ and (c) single crystal (001)

CONCLUSIONS

Alumina and titania have been studied as model dielectrics. The behaviour of these materials has revealed important general observations regarding the influence of processing and microstructure on the dielectric properties of sintered ceramics. In both alumina and titania porosity was seen to affect the dielectric constant as might be expected. The influence of porosity on the tan δ of alumina is profound and even small amounts of porosity can cause a dramatic increase in the tan δ. In titania ceramics the influence of porosity was also profound but for different reasons. Here, the closing of interlinked pores as densification proceeded caused a loss of oxygen in the final dense ceramic and this was associated with a very poor Q. To overcome this it was found that the addition of an oxygen ion conductor (YSZ) enabled better oxygenation of the titania and hence an improvement in Q at near full density. The reasons for the improvement are not clear at present but it is possible that the oxygen ion conduction properties of the YSZ are fortuitous and that the presence of 3+ ions are a critical feature.

ACKNOWLEDGEMENTS

This work is partially supported by the EPSRC Grant No. GR/K70649 and by the European Community Brite-Euram project DiHiMiCo (BE1824).

REFERENCES

1 Y. KONISHI "Novel dielectric waveguide components - microwave applications of new ceramic materials" *Proc. IEEE* **79** (6)726-740 (1991)

2. D. KAJFEZ and P. GUILLON, Dielectric Resonators, (Artech House, Zurich, 1986)

3 N.McN. ALFORD and S.J. PENN."Sintered alumina with low dielectric loss" J. Appl. Phys., **80** 5895-5898 (1996)

4. S. PENN, N. McN. ALFORD, A. TEMPLETON, M. XU, X. WANG, M. REECE, K. SCHRAPEL, "Effect of Porosity and Grain Size on the Dielectric Properties of Sintered Alumina", J. Am. Ceram. Soc. **80** p. 1885-1888 (1997)

5 V. B. BRAGINSKY V. S. ILCHENKO and K. S. BAGDASSAROV. "Experimental observation of fundamental microwave absorption in high quality dielectric crystals". Physics Letters A **120** p300-305 (1987)

6 H. O'BRYAN and J. THOMSON. "$Ba_2Ti_9O_{20}$ Phase equilibria" *J. Amer. Ceram. Soc* **66**(1) 66-68 (1983)

7 H.M. .O'BRYAN, J THOMSON and J.K. PLOURDE. "A new BaO-TiO_2 compound with temperature-stable high permittivity and low microwave loss" *J. Am. Ceram. Soc* **57** (10) 450-453 (1974)

8 H. O'BRYAN "Identification of surface phases on $BaTiO_3$-TiO_2 ceramics" Am. Ceram. Soc. Bull **66**(4) 677-80 (1987)

9 T. NEGAS, G YEAGER, S BELL, N. COATS and I. MINIS. "$BaTi_4O_9$/$Ba_2Ti_9O_{20}$ - based ceramics resurrected for modern microwave applications. *Am. Ceram. Soc. Bull.* **72**(1) 80-89 (1993)

10 W.D. KINGERY, H.K. BOWEN and D.R. UHLMANN. Introduction to Ceramics 2nd edition (Wiley, New York 1976)

11 V.L.GUREVICH and A.K.TAGANTSEV. *Advances in Physics* **40** (6) 719-767 (1991)

RELIABILITY OF NEW PACKAGING CONCEPTS

PROF. NIHAL SINNADURAI
Principal Consultant
Advanced Materials & Processes Department
TWI, Abington, Cambridge, CB1 6AL, UK
'Phone: (+44) (0) 1223 891162, 'Fax: (+44) (0) 1223 892588

ABSTRACT

Commercial pressures require that modern microelectronics deliver more complexity in less space and more reliability at lower cost. Today, most of the microelectronics packaging needs are met by semiconductor devices in plastic surface mount (SM) packages. Microelectronics packaging of the future will be either bare chip or chip size/scale packaging (CSP). Of the 40 billion SM packaged ICs manufactured in 1998, CSPs will be a small 1.1% but growing at 62% (compound annual growth rate (CAGR). The use of direct bonded chip-on-board (COB) and flip chip (FC) technology for custom solutions may match the growth of CSPs. The popcorn problem of existing plastic packages is solved. The problem which was more severe with the smaller and thinner encapsulations of CSPs, is also solved with modern hydrophobic materials and new package designs. The portents are promising for reliable CSP and COB solutions.

1. INTRODUCTION

Modern microelectronics are required by the marketplace to continue to deliver more for less: - more complexity in less space, and more reliability at lower cost. Moreover, future applications environments are likely to be more hostile as electronics increases its penetration: in automotives (Delco expects 50% of the value of the car to be in the electronics by 2002 and Volvo is introducing the Car Area Network with a HC08 microcontroller), more fly-by-wire aircraft, more mobile telephony, greater penetration of electronics in the developing world (which also has the more hostile climates[1]), increasing medical electronics implants (the bionic person is not far off[2]). Recent innovations in electronics packaging materials and technologies are enabling greater product miniaturisation whilst increasing manufacturability, testability and robustness.

2. MARKETS

Today, most of the microelectronics packaging needs are met by semiconductor devices in plastic surface mount packages[3] (Figure 1), and surface mount technology (SMT) is the dominant electronics assembly technology . Microelectronics packaging options to meet the future demands will be either not to package (bare chip technologies such as chip-on-board (COB) and flip chip (FC)) or to use chip size/scale packaging (CSP). A useful road-map of microelectronics packaging is shown in Figure 2[4]. Some 40 billion SM packaged ICs will be manufactured in 1998[3]. The quantity of CSPs will be a small 1.1% of that market, but growing at 62% CAGR[5] and manufacturing efficiency will benefit from volume related process controls and consequent improvements in quality and associated reliability. Direct bonded COB and FC technologies are likely to have a slower market growth than that of CSPs and are more suited to custom solutions which have lower yield assembly processes.

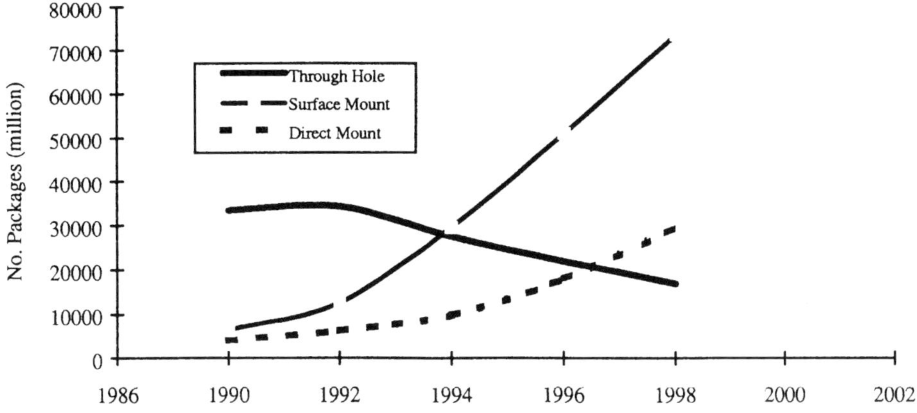

Figure 1. Semiconductor Device Package Market Trends

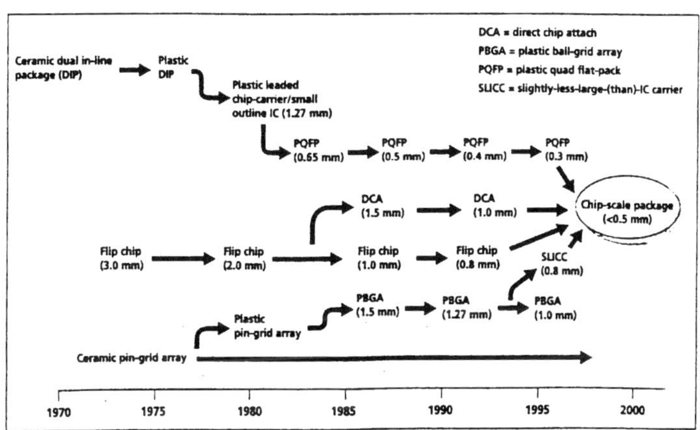

Figure 2. Evolution of Microelectronics Packaging Towards Chip Scale Packaging

3. RELIABILITY DEVELOPMENT

As COB and many CSP types are already in use and others are being developed, the question for the near future is whether the high reliability benefits of present commercial packages can be carried forward and improved or whether potential hazards will be exacerbated. Today, reliability is well proven for larger scale packaging. The significant assembly problem observed when SMT packages were wave or reflow soldered was "Popcorning" which is the explosive expansion of moisture absorbed into large delaminations beneath the die paddle (the metal flag to which the IC die is attached). Early observations were that popcorn damage, was more severe with the smaller and thinner encapsulations of CSPs because of the greater opportunities for moisture penetration and retention beneath larger chip sizes and greater fragility of the thinner packages[6]. Packages made with modern materials can contain about 0.3% by volume of micro-voids[7], but are readily dried out and do not result in damage during solder reflow[7]. Popcorning is a problem only during initial assembly and is now managed and avoided by good manufacturing practice. Solutions to avoid popcorning include new hydrophobic encapsulating materials, new package designs and pre-conditioning of the packages prior to assembly. Popcorning is not a reliability hazard.

Solid moulded packages can contain small delaminations (Figure 3)[8] because stress is relieved following the high temperature and pressure of transfer moulding. The density and location of delaminations can be revealed by the use of C-SAM (C-mode scanning acoustic microscopy). Intelligent interpretation is important because thresholds can be set to show total or no delaminations in the same package. The concern had been that delamination sites would store and release contaminants or lead to package fracture by rapid expansion of stored moisture. However, it is not the quantity but the location of delaminations that matter, because plastic encapsulated device reliability is safeguarded by achieving a good interface bond between device chip and the polymer[9] (Figure 4). Plastic encapsulation by silicone applied to a clean silicon IC surface can safeguard the reliability of the IC for more than 100 years in harsh tropical climates[1,9] and there are more than 100 million device hours of reliability testing to prove the excellent reliability of plastic encapsulated semiconductor devices[1] and optoelectronics devices[10]. An additional benefit from such encapsulations is that they absorb alpha particles, thereby protecting the semiconductor device from induced soft errors. Reliability studies involving long term damp and dry heat stress studies have also shown there is no direct correlation between moisture absorption or delaminations and device failure[8,9].

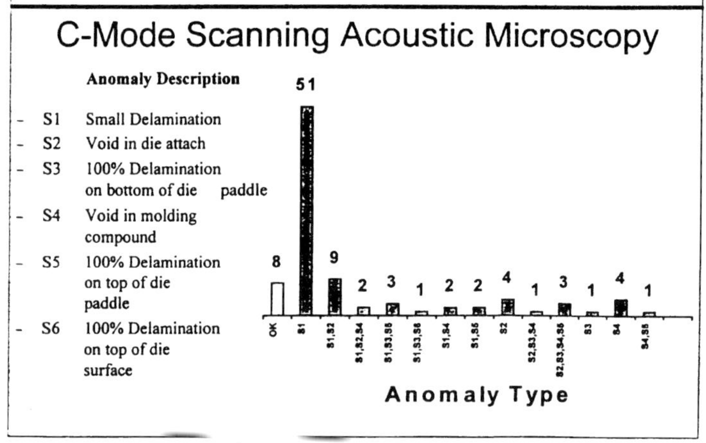

Figure 3. Delamination Distribution in a Population of Plastic Packaged ICs[8]

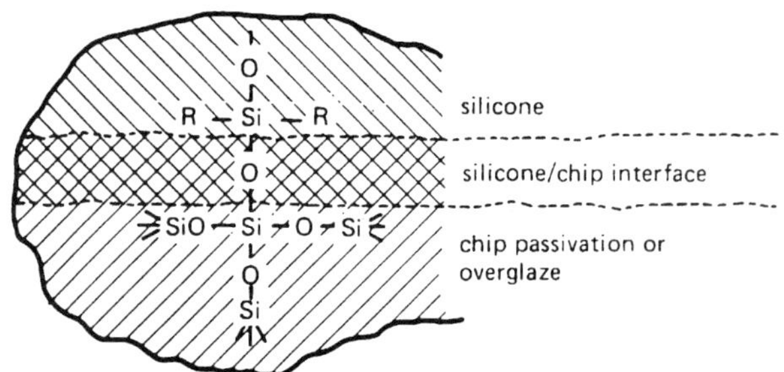

Figure 4. Chemisorption of Silicone encapsulation to Silicon Passivation Surface[9]

3.1 Popcorn-Proof Packaging Solutions

In addition to engineering the plastic packaging process to achieve a good plastic-to-chip bond, precautions are taken at solder assembly to ensure that moisture is excluded from the packages to avoid popcorning. Such precautions and the characterisation of moisture retention properties of encapsulation materials are specified in JEDEC JESD22-A112-A[11] from which the definitions of moisture resistance levels are reproduced in Table I.

Table I. Moisture Sensitivity Levels of PEDs According to JEDEC JESD22-A112-A

<table>
<tr><th>LEVEL</th><th colspan="2">FLOOR LIFE</th><th colspan="4">SOAK REQUIREMENTS
(for preconditioning)</th></tr>
<tr><th></th><th>Conditions</th><th>Time</th><th colspan="3">Time</th><th>Conditions</th></tr>
<tr><td>1</td><td>≤30°C/90%RH</td><td>Unlimited</td><td colspan="3">168 hours</td><td>85°C/85%RH</td></tr>
<tr><td>2</td><td>≤30°C/60%RH</td><td>1 year</td><td colspan="3">168 hours</td><td>85°C/60%RH</td></tr>
<tr><td></td><td></td><td></td><td colspan="3">Time (hours)</td><td></td></tr>
<tr><td></td><td></td><td></td><td>x +</td><td>y =</td><td>z</td><td></td></tr>
<tr><td>3</td><td>≤30°C/60%RH</td><td>168 hours</td><td>24</td><td>168</td><td>192</td><td>30°C/60%RH</td></tr>
<tr><td>4</td><td>≤30°C/60%RH</td><td>72 hours</td><td>24</td><td>72</td><td>96</td><td>30°C/60%RH</td></tr>
<tr><td>5</td><td>≤30°C/60%RH</td><td>24 hours</td><td>24</td><td>24/48</td><td>48/72</td><td>30°C/60%RH</td></tr>
<tr><td>6</td><td>≤30°C/60%RH</td><td>6 hours</td><td>0</td><td>6</td><td>6</td><td>30°C/60%RH</td></tr>
</table>

x = Default value of the semiconductor manufacturers time between bake and bag plus the maximum time allowed out of the bag at the distributors facility. The actual times may be used rather than the default times, but they must be used if they exceed the default times.

y = Floor life of the package after it is removed from dry pack bag.

z = Total soak time for evaluation.

Popcorning problems have been overcome not only by improved materials, but also by new developments of CSP. The lead-on-chip (LOC) CSP[12] has leads glued directly to the centre of the IC chip and eliminates plastic moulding on the underside of the chip (Figure 5). The chip-side-support package (CSSP)[13] has eliminated the die paddle (Figure 6).

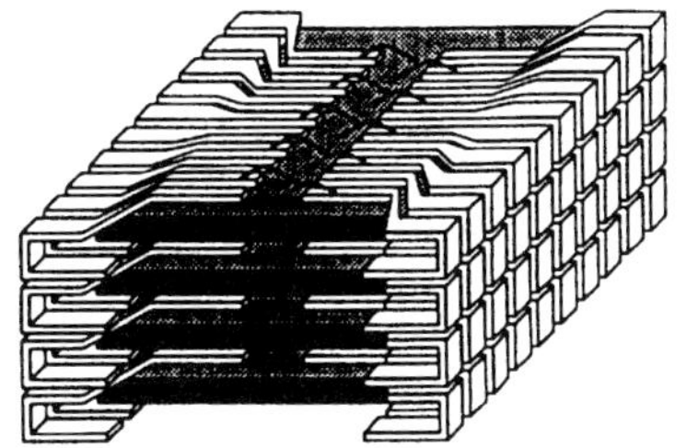

Figure 5. Stacked LOC CSPs without die-paddle or moulding under the chip

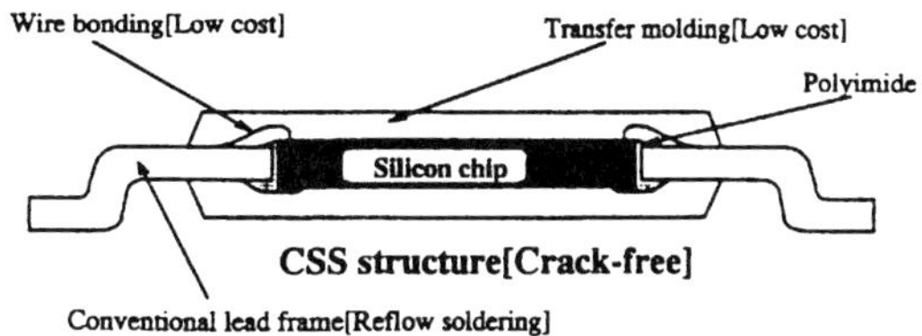

Figure 6. CSS structure without die- paddle but with moulding under chip

3.2 Reliable Encapsulation

The reliability challenges that faced micropackages originally stemmed from the Military Standards (MIL STDs) which included an outright ban on plastic packages. Meanwhile the rest of the world improved commercial plastic encapsulations and manufacturing methods and produced solutions for industrial high reliability applications. The new enlightened MIL and space approach is to take advantage of the quality, reliability and cost benefits obtainable from microelectronics manufactured for the high volume markets (referred to as "commercial" markets). Today's plastic packaged microelectronics devices achieve quality yields at ppm (parts per million) defect levels, and operating reliability of the order of 10 FITS (failure units = 1 in 10^9 device hours) because they are manufactured using vastly improved materials compared with those which led to the MIL ban some 30 years ago. The PURE (Plastic encapsulated devices (PEDs) Used in Ruggedised Environments) Association comprising a Franco-Swedish collaboration[14], has developed qualification procedures for PEDs for military applications. The PURE association carries out its own audits, which are made available to their equipment designers. As a result, companies which are part of the PURE association use substantial proportions of PEDs in their equipment.

Plastic encapsulation technology has improved dramatically over the past two decades, aided by effective materials assessments[15], and has resulted in high reliability standard commercial plastic packaging suitable also for applications in harsh climates[1]. The strong evidence derived from these developments yielded the authentic S-H model for the temperature-humidity-bias (THB) mechanisms of failure and accelerated ageing of electronics components[16].

3.3. Joint Reliability of Flip COB and Flip CSP

While there is now considerable confidence in the protection afforded by encapsulation of large microelectronics packages and CSPs, the reliability issue receiving significant attention is the failure mechanisms associated with fatigue in the fine pitched joints of FC COB (FCOB) and CSPs. Fatigue can be induced in the joints by cyclic strains due to temperature excursions. Reliability studies have shown that fatigue failures will decrease if the space between the FC or flipped ball grid array (BGA) is filled with an adhesive/encapsulant which supports the solder bump joints and mechanically couples the IC and Printed Circuit Board (PCB) [17]. The benefits come from matching the coefficient of thermal expansion (CTE) of the encapsulant with the solder, the PCB and the chip. Optimum survival is obtained by matching with the solder and can result in doubling the strain cycles to failure compared with no underfill, e.g. increase from 6000 cycles to 10000 cycles 0°C to 100°C. In order to provide the necessary mechanical coupling, it is essential for the encapsulant to reach and wet all the interfaces in the gap - solder mask, solder bumps and chip passivation. A lack of wetting or failure to penetrate will result in voids and cause excessive shear forces on the joints. C-SAM analyses are useful to reveal voids around solder bumps (Figures 7 and 8) and failure analysis by micro-sectioning has shown that solder is extruded into voids adjacent to the bumps, thus permitting earlier failure at these sites.

CSP packaged microcircuits also benefit from matching the package CTE to that of the underlying PCB, thereby minimising strain during temperature excursions. Power Tape Ball Grid Array (TBGA) CSPs developed with embedded copper heat sink stiffeners with a CTE of 16.6 ppm/°C which match the CTE of the PCB. These TBGAs soldered to FR4 PCBs outlasted more conventional BGAs soldered to the same FR4 PCBs (Figure 9).

The portents are promising for reliable CSP and COB solutions.

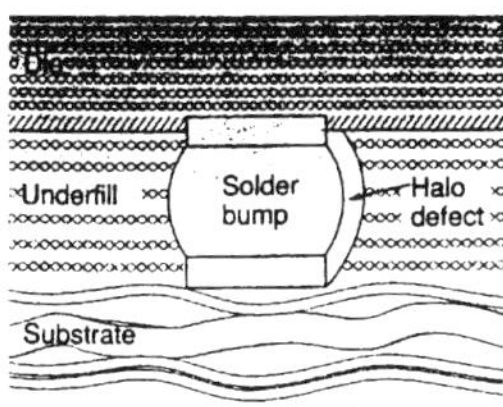

Figure 7. Voids around bumps appear as 'haloes' in CSAM images

Figure 8. Schematic section showing space for bump to move

(Illustrations courtesy of Sonoscan)

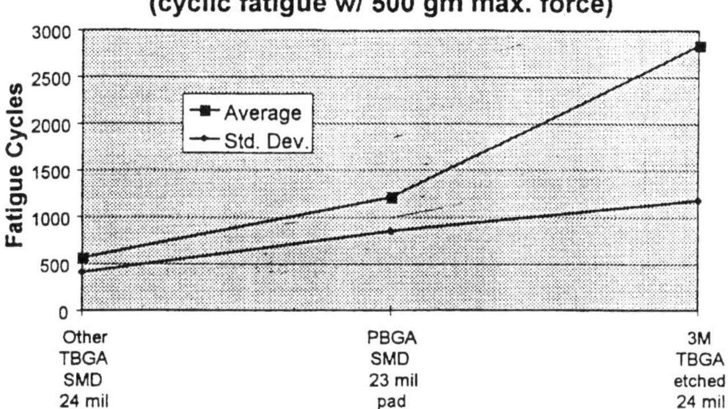

Figure 9. Fatigue Testing of PBGA compared with other TBGA and Plastic BGA.

REFERENCES

1. N.SINNADURAI: 'Plastic Packaging is Highly Reliable', *IEEE Transactions on Reliability*, June 1996, **45** (2), 184-193.
2. N DONALDSON: 'Developments in Medical Electronics', *Proceedings of IMAPS-UK 30th Anniversary Conference 'Electronics - The Next Decade'*, IMAPS-UK, 1-2 June '98. Also 'On My Feet Again', (Referring to pioneering work by Dr N Donaldson), *BBC1 Documentary*, 1 February 1996.
3. N.SINNADURAI: 'Microelectronics Markets in Europe', *Proceedings of the 10th European Microelectronics Conference*, ISHM-Europe, Copenhagen, May 1995.
4. P.THOMPSON: 'Chip Scale Packaging', *IEEE Spectrum, August 1997*, 36-43.
5. 'The Changing Assembly Picture', *Surface Mount Technology*, IHS Publishing Group, August 1997, 32-46,
6. H BERG, G GANESAN AND G LEWIS: 'Popcorn-proof Packaging', *Advanced Packaging*, May/June 1996, 18-22,.
7. M A SCHEN, W WU, W E WALLACE, N BECK-TAN, D VANDERHART AND G T DAVIS, 'Molecular Insights on Interfacial Properties and Moisture Uptake of Plastic Packaging Materials', *Advancing Microelectronics*, July/August 1997, 26-27.
8. A CASASNOVAS AND J W WHITE: 'The Effects of Long Term Dormant Storage on Plastic Encapsulated Microcircuits', *Proc. 3rd ESA Electronics Components Conference*, Noordwijk, April 1997, 365-374.
9. F N SINNADURAI: '*Handbook of Microelectronics Packaging and Interconnection Technologies*', Electrochemical Publications, Isle of Man, 1985.
10. I P HALL 'Novel Techniques for Low Cost, High Performance, Optoelectronic Component Assembly', *Hybrid Circuits*, June 1991, **34** 352-365.
11. 'Moisture-Induced Stress Sensitivity for Plastic Surface Mount Devices', *Specification JESD22-A112-A*, Joint Electron Devices Engineering Council (JEDEC).
12. N TAKETANI, K HATANO, H SUGIMOTO, O YOSHIOKA and G MURAKAMI, 'CSP with LOC Technology', *International Journal of Microcircuits and Electronic Packaging*, Q2 1997, **20** (2) 96-101.
13. T NAKAZAWA, Y INOUE, K SAWADA AND T SUDO: 'A novel Structure to Realise Crack-Free Plastic Packages During the Reflow Soldering Process - Development of Chip Side Support Package (CSSP)', *IEEE Trans. Components, Packaging and Manufacturing Technology - Part C*, January 1996, **19** (1) 61-69.
14. 'Plastic Component Programme (PCP) Qualification Reference Document (QRD)', *French MOD Documents*, **SCTF/96.0211** - Summary.
15. 'Type Approval of Encapsulating Resins, Coatings and Adhesive Materials for Semiconductor Devices and Microcircuits', *British Telecom Specification M219F*, BT Procurement Executive, Materials & Science Section QA6.4.2, March 1982.
16. N. SINNADURAI, 'Proving and Improving Electronics Reliability', *TWI Bulletin*, September/October 1997, 91-94.
17. D GAMOTA and C MELTON, 'Advanced Encapsulant Materials Systems for Flip Chip', *Advancing Microelectronics*, July/August 1997, 22-24
18. R D SCHUELLER, D AESCHLIMAN and C T HAN, 'Performance and Reliability of a Cavity-Down Tape BGA Package', Proc. EMIT'98, IMAPS-India, February 1998, 233-245

Index